Parasitology Research Monographs

Volume 4

Series Editor:
Heinz Mehlhorn
Department of Zoomorphology
Cell Biology and Parasitology
Heinrich Heine University
Universitätsstrasse 1
40225 Düsseldorf
Germany

For further volumes:
http://www.springer.com/series/8816

Heinz Mehlhorn • Kevin S.W. Tan •
Hisao Yoshikawa
Editors

Blastocystis: Pathogen or Passenger?

An Evaluation of 101 Years of Research

Springer

Editors
Heinz Mehlhorn
Department of Zoomorphology
Cell Biology and Parasitology
Heinrich Heine University
Düsseldorf
Germany

Kevin S.W. Tan
Laboratory of Molecular and Cellular Parasitology
Department of Microbiology
National University of Singapore
Singapore

Hisao Yoshikawa
Department of Biological Sciences
Nara Women's University
Nara
Japan

ISSN 2192-3671 ISSN 2192-368X (electronic)
ISBN 978-3-642-43306-1 ISBN 978-3-642-32738-4 (eBook)
DOI 10.1007/978-3-642-32738-4
Springer Heidelberg New York Dordrecht London

Printed on acid-free paper

Springer is part of Springer Science+Business Media (www.springer.com)

Preface

For about 100 years *Blastocystis* stages are known to occur in normal and diarrheic feces of humans and of many species of animals. Many important papers in this daily enlarging field of research appeared in Parasitology Research at Springer, Heidelberg, New York. From the very beginning of the research approaches, there had been intense disputes whether there exist different species in humans (e.g., *B. hominis*) and in animals, whether these cyst-like appearing spherical stages are true pathogens or just live as passaging commensals, whether they belong to the phyla of fungi or protozoans, whether they induce diarrheas in humans and animals or whether they reproduce better in hosts under diarrheic conditions, and finally whether they are really able to act as zoonotic organisms switching from humans to animals and back or take the inverse route of primary infection.

Especially the last 45 years delivered worldwide thousands of publications on *Blastocystis* presenting results that were obtained by the use of a wide spectrum of standard and of very sophisticated methods. Although these provide deep insights into the life cycle, the physiology, morphology, and genetics of this peculiar organism remained a mysterious sphinx among the intestine-inhabiting organisms.

Against this background this volume of Parasitology Research Monographs unites renowned specialists and asked them to "put all hard facts onto the table" that had been obtained in *Blastocystis* research as well as in the knowledge on diarrheas and on other apparently zoonotic agents of disease.

Thus this volume tries to evaluate the recent knowledge in the *Blastocystis* field. When focussing on open questions the book may stimulate relevant research in burning problems such as adequate diagnosis and chemotherapy. Its detailed review articles are directed not only at researchers in this field and physicians and veterinarians who are confronted with infected individuals or animals but also to teachers, students, and technical staff in the fields of microbiology, parasitology, and diagnostic methods.

September 2012

Heinz Mehlhorn
Kevin S.W. Tan
Hisao Yoshikawa

Acknowledgment

The quick, careful, and attractive publication of so much data based on own research and/or selected from papers of a very broad spectrum of international authors is not possible without the help of many persons. At first we thank all contributors for their in-time delivery of the manuscripts, so that latest research aspects could have been included. Then we are indebted to the efforts of all coworkers of each group who gave text and figures their final shape. Our thanks are also directed to Mrs. Anette Lindqvist from Springer and the production team of SPi Global. Their duly and competent efforts made it possible to present these comprehensive insights into the present knowledge of an organism, the importance of which is surely still underestimated despite its common and worldwide presence.

September 2012

Heinz Mehlhorn
Kevin S.W. Tan
Hisao Yoshikawa

Contents

About the Authors

Prof. Dr. Heinz Mehlhorn, Düsseldorf, Germany. He has investigated the transmission pathways of human and animal parasites for over 40 years and his university spin-off company Alpha-Biocare has developed various antiparasitic medical products based on more than 20 patents. He has published 25 books, more than 250 original papers, and has served as Managing Editor of the journal Parasitology Research since 1981.

Prof. Dr. Kevin SW Tan, Singapore, a former postdoctoral fellow at the Rockefeller University, is a specialist in molecular and cellular parasitology research in the fields of *Blastocystis* and *Plasmodium falciparum*, and he has received various national and international awards as well as considerable research grants for his work to date. More recently, he has funneled the research output of his many publications into the foundation of the company Biolynx Technology.

Prof. Dr. Hisao Yoshikawa, Nara, Japan. He has been an active researcher in many fields of medical and zoological parasitology for over 30 years. Over the course of this time, he has intensively studied the ultrastructure, cell biology, and molecular phylogeny of many isolates of *Blastocystis*.

Chapter 1
Landmarks in the Discovery of *Blastocystis* Stages

Heinz Mehlhorn, Minoru Yamada, Hisao Yoshikawa, Kevin S.W. Tan, Haris Mirza, and Kenneth Boorom

Abstract This chapter represents a list of papers that were considered by the authors of this book as important "breakthroughs" in the knowledge of the different features in the history of the *Blastocystis* research during the last 101 years. Of course this is a selection, but due to space, others were not considered in this chapter, but belong to the basic knowledge collected in the other chapters of this book and are cited at the end of each chapter.

The first observations of Alexeieff (1911) and Brumpt (1912) of stages, which looked different from other intestinal spherical, cyst-like, or similar appearing

---

H. Mehlhorn (✉)
Institute for Parasitology, Heinrich Heine University, Universitätsstr. 1,
40225 Düsseldorf, Germany
e-mail: mehlhorn@uni-duesseldorf.de

M. Yamada
Department of Infectious Diseases, Kyoto Prefectural University of Medicine, Kyoto, Japan

H. Yoshikawa
Department of Biological Sciences, Nara Women's University, Nara, Japan
e-mail: h.yoshikawa@cc.nara-wu.ac.jp

K.S.W. Tan
Laboratory of Molecular and Cellular Parasitology, Department of Microbiology,
Yong Loo Lin School of Medicine, National University of Singapore, 5 Science Drive 2,
Singapore 117597, Singapore
e-mail: mictank@nus.edu.sg

H. Mirza
Yale University School of Medicine, 333 Cedar Street, New Haven, CT 06510, USA
e-mail: Haris.mirza@yale.edu; Drmirza06@yahoo.com

K. Boorom
Blastocystis Research Foundation, 5060 SW Philomath Blvd, #202, Corvallis, OR 97333, USA
e-mail: kboorom@bhomcenter.org

H. Mehlhorn et al. (eds.), *Blastocystis: Pathogen or Passenger?*,
Parasitology Research Monographs 4,
DOI 10.1007/978-3-642-32738-4_1,

unicellular organisms inside the intestinal fluid of humans, were done by means of light microscopy. This method, although it had opened new and deeper insights after the amelioration of the microscopical techniques by Carl Zeiss and Ernst Abbe and after development of different coloration methods (especially for blood smears) by Gustav Giemsa, Wilhelm Schüffner or Georg Maurer, did not allow to differentiate clearly between the various unicellular organisms, that occur regularly or occasionally in the intestine of humans and animals. Especially the newly described organism called *Blastocystis hominis* by Brumpt (1912) could not be easily classified when using early microscopical techniques. Thus morphological and some early physiological criteria led to classifications either to the group of protozoans or fungi. Even electron microscopical investigations (e.g. Zierdt and Tan 1976; Mehlhorn 1983; Figs. 1.1 and 1.2) did not solve the problems, which are even today under discussion, although especially the studies since 1967 brought enormous progress. Therefore the following list presents some publications which are considered as "landmarks" in the history of the discovery of the "*Blastocystis* features" by the authors of this book. Of course this list represents a selection, which is not complete and it is clear that there exist further excellent, highly valuable research results by a large number of renowned groups. We apologize for not naming them. However, all those researchers contributed significantly to the present stage of knowledge in this field.

1911 and 1912	First descriptions of *Blastocystis* were given by Alexeieff (1911) and Brumpt (1912). The term *Blastocystis* was coined by Alexieff since the envelope of the parasite reminded him at the capsule of the fungus *Blastomyces* and because the parasite formed cysts (*Greek*: kystis = bladder, cyst). It was later appreciated that the cysts of the parasite were not true cysts but very likely what is currently termed multivacuolar forms, where the periphery showed a somewhat-condensed cytoplasm. Alexeieff called the organism *Blastocystis enterocola* and a year later, Brumpt coined the term *Blastocystis hominis*. The latter term is now widely used. Thus the organism was initially classified as a yeast.
1967	Zierdt et al. (1967) pointed out the protozoan organization of the *Blastocystis* stages and reclassified them as members of the group of Protozoa.
1973	Zierdt and colleagues intensified morphological studies on *Blastocystis* (Zierdt 1973; Tan and Zierdt 1973).
1976	Description of the pathogenic potential of *Blastocystis* (Phillips and Zierdt 1976).
1983	First approaches to treatment trials in *Blastocystis hominis* (Zierdt et al. 1983).
1986	First study to recommend that *Blastocystis* patients be diagnosed with IBS (Markell and Udkow 1986).
1987	Description of morphology of animal species/strains of *Blastocystis* (Yamada et al. 1987).

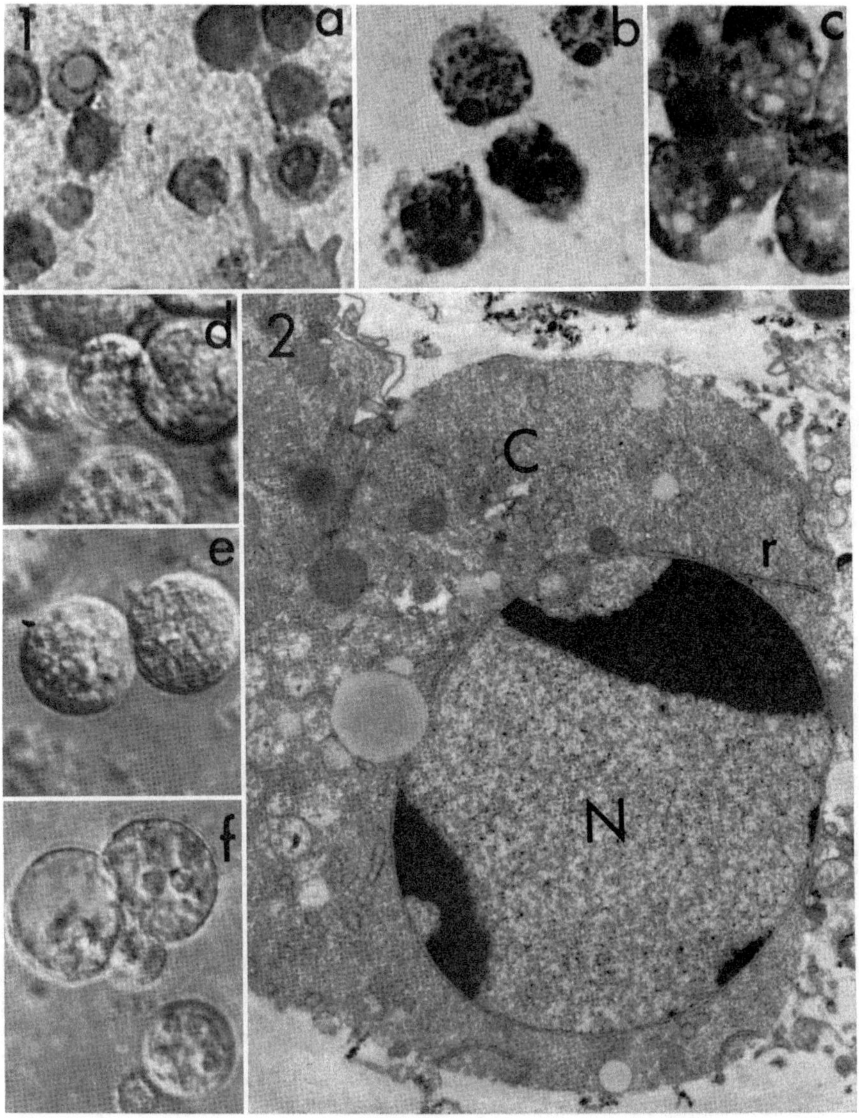

Fig. 1.1 It originates from the paper of Zierdt CH and Tan HK (1976) Ultrastructure and light microscopic appearance of *Blastocystis hominis* in a patient with enteritic disease. Z Parasitenkd (now Parasitol Res) 50:277–283. **Part 1a–c** Hematoxylin and eosin-stained sections. Note *light* and *dark spherical granules* within organisms. Nuclei cannot be clearly identified. **Part 1d–f** Nomarski interference contrast photomicrographs showing vacuolar and granular structures within organisms. **Part 2** Electron micrograph of a section of an organism. A nucleus (*N*) with a crescent-like concentration of nuclear chromatin and cytoplasm (*C*) with vesicles and granules are evident. Note segments of endoplasmatic reticulum (*r*) near cell surface

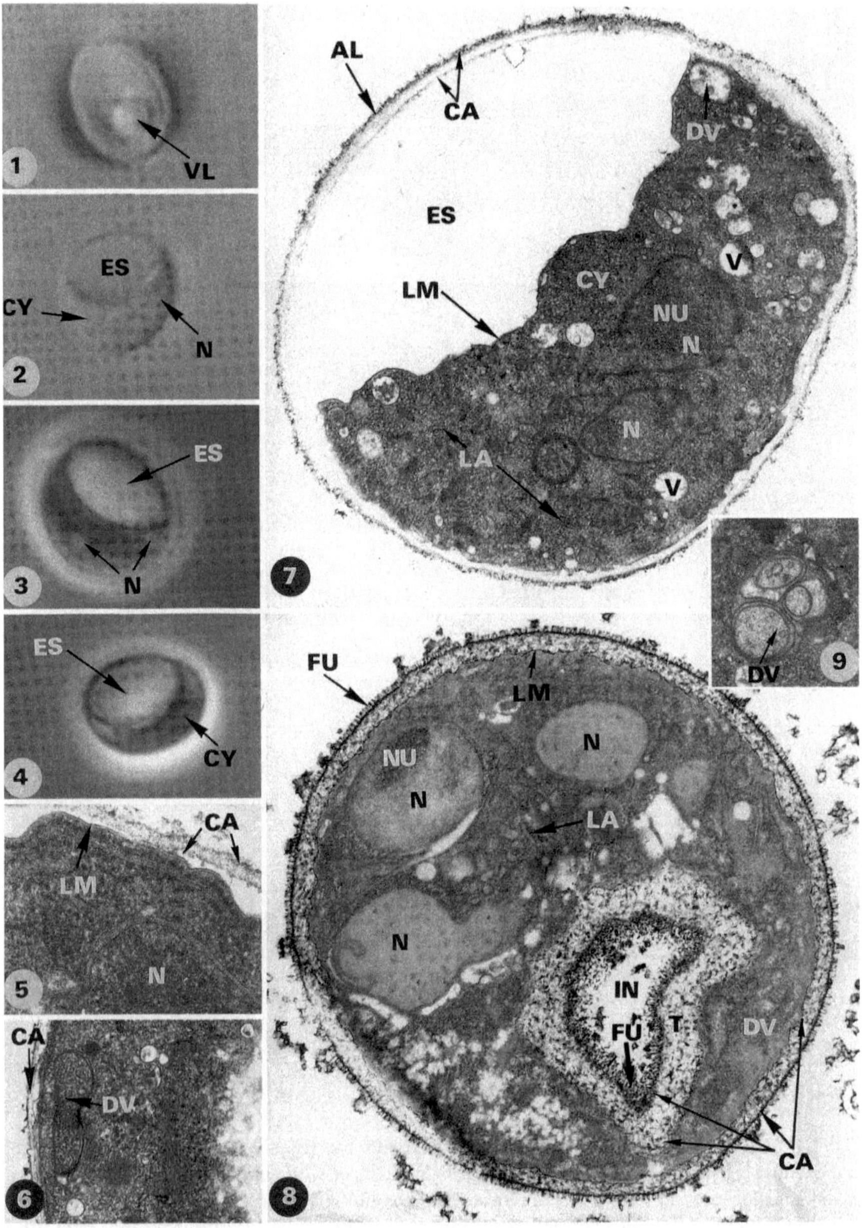

Fig. 1.2 It originates from the paper of Mehlhorn H (1983) *Blastocystis hominis*, Brumpt 1912: Are there different stages or species? Parasitol Res 74:393–395. **Part 1–9** *Blastocystis*-like stages from human diarrheal stools. 1–4 Light micrographs. Note the *different shape* of the cytoplasm. X2820. 5–9 Electron micrographs. Two types of cyst-like organisms are found; type 1 (5–7, 9) has nongranulated capsule (*CA*), which is different from that in type 2 (8). The cyst-like stage in **Part 8** is cut tangentially through the cytoplasm, hence the empty space (*ES*) is

1993	First detailed study on a *Blastocystis* epidemic (Nimri 1993).
1996	Intensification of taxonomic approaches (Nakamura et al. 1996).
1996	By phylogenetic analysis of the ssrRNA genes, *Blastocystis* is reclassified as a stramenopile. The stramenopiles, synonymous with *Heterokonta* and *Chromista*, are a complex collection of "botanical" protists comprising heterotrophic and photosynthetic representatives (Silberman et al. 1996).
1996	A method for the in vitro clonal growth of the parasite is reported (Tan et al. 1996).
1997	First mouse model in *Blastocystis* research (Moe et al. 1997).
1997	First studies to suggest that *Blastocystis* infection causes irritable bowel syndrome (Hussain et al. 1997).
1999	First studies to suggest that some *Blastocystis* isolates are associated with acute infections, while others may be associated with chronic infections (Lanuza et al. 1999).
2000	First long-term study (2 years) on *Blastocystis* immune response in patients with and without gastrointestinal symptoms (Kaneda et al. 2000).
2002	First study to report that common clinical diagnostic techniques fail to identify most *Blastocystis* infections (Leelayoova et al. 2002).
2003	First placebo-controlled trial for treatment of *Blastocystis* infection with antimicrobials (Nigro et al. 2003).
2003	First study to report a difference in *Blastocystis* antigen immune response between symptomatic and asymptomatic patients (Mahmoud and Saleh 2003).
2005	First major analysis of *Blastocystis* genotypes from multiple countries (Noel et al. 2005).
2005	First study to report death in animal models following experimental infection with *Blastocystis* (Yao et al. 2005).
2005	First large epidemiological study on *Blastocystis* infection (Amin 2006).
2007	First investigation of waterborne transmission of *Blastocystis* including molecular epidemiology and large-scale population study of *Blastocystis* (Li et al. 2007a,b).
2007	A new terminology of *Blastocystis* is proposed, which is today widely accepted (Stensvold et al. 2007).
2008	Study indicating that *Blastocystis* transmission may occur by drinking water (Leelayoova et al. 2008).

Fig. 1.2 (continued) not visible. *AL* amorphous outer layer (by fecal contamination?); *CA* capsule; *CY* cytoplasm; *DV* double-walled vacuole (mitochondrion?); *ES* empty space; *FU* fuzzy regular outer layer of the capsule; *IN* invagination surface depression; *LA* lacunes, endoplasmatic reticulum-like structures; *LM* limiting membrane of the cytoplasm; *N* nucleus; *NU* nucleolar granules; *T* tangentially sectioned capsule; *V* vacuole; *VL* vacuole-like structure

2008	First study to perform experimental animal infection with multiple *Blastocystis* isolated that were subtyped (Hussein et al. 2008).
2008	First real-time PCR test for *Blastocystis* infection (Jones li et al. 2008).
2008	First systematic review of *Blastocystis* literature (Boorom et al. 2008).
2009	First genotyping of *Blastocystis* infection in cancer patients (Tan et al. 2009).
2010	First study to report multiple immunosuppressive properties of *Blastocystis* secretory substances (Chandramathi et al. 2010).
2010	Legumain, the first virulence factor of *Blastocystis* is cloned and characterized (Wu et al. 2010).
2011	First complete genome sequencing of *Blastocystis* (Denoeud et al. 2011).
2011	First study to report that DNA extraction kits have greatly varying performance level (Yoshikawa et al. 2011).
2011	First study to report that *Blastocystis* downregulated nitric oxide production (Mirza et al. 2011a,b).

References

Alexeieff A (1911) Sur la nature des formations dites kystes de *Trichomonas intestinalis*. C R Soc Biol 71:296–298

Amin O (2006) Epidemiology of *Blastocystis hominis* in the United States. Res J Parasitol (2005) 1:1–10

Boorom KF, Smith H, Nimri L, Viscogliosi E, Spanakos G, Parkar U, Li LH, Zhou XN, Ok UZ, Leelayoova S, Jones MS (2008) Oh my aching gut: irritable bowel syndrome, *Blastocystis*, and asymptomatic infection. Parasit Vectors 1:40–60

Brumpt E (1912) *Blastocystis hominis* n sp. et formes voisines. Bull Soc Pathol Exot 5:725–730

Chandramathi S, Suresh K, Kuppusamy UR (2010) Solubilized antigen of *Blastocystis hominis* facilitates the growth of human colorectal cancer cells, HCT116. Parasitol Res 106:941–945

Denoeud F, Roussel M, Noel B, Wawrzyniak I, Da Silva C, Diogon M, Viscogliosi E, Brochier-Armanet C, Couloux A, Poulain J, Segurens B, Anthouard V, Texier C, Blot N, Poirier P, Ng GC, Tan KSW, Artiguenave F, Jaillon O, Aury JM, Delbac F, Wincker P, Vivarès CP, El Alaoui H (2011) Genome sequence of the stramenopile *Blastocystis*, a human anaerobic parasite. Genome Biol 12:R29

Hussain R, Jaferi W, Zuberi S, Baqai R, Abrar N, Ahmed A, Zaman V (1997) Significantly increased IgG2 subclass antibody levels to *Blastocystis hominis* in patients with irritable bowel syndrome. Am J Trop Med Hyg 56:301–306

Hussein EM, Hussein AM, Eida MM, Atwa MM (2008) Pathophysiological variability of different genotypes of human *Blastocystis* hominis Egyptian isolates in experimentally infected rats. Parasitol Res 102:853–860

Jones Ii MS, Ganac RD, Hiser G, Hudson NR, Le A, Whipps CM (2008) Detection of *Blastocystis* from stool samples using real-time PCR. Parasitol Res 104:341–345

Kaneda Y, Horiki N, Cheng X, Tachibana H, Tsutsumi Y (2000) Serologic response to *Blastocystis hominis* infection in asymptomatic individuals. Tokai J Exp Clin Med 25:51–56

Lanuza MD, Carbajal JA, Villar J, Mir A, Borras R (1999) Soluble-protein and antigenic heterogeneity in axenic *Blastocystis hominis* isolates: pathogenic implications. Parasitol Res 85:93–97

Leelayoova S, Siripattanapipong S, Thathaisong U, Naaglor T, Taamasri P, Piyaraj P, Mungthin M (2008) Drinking water: a possible source of *Blastocystis* spp. subtype 1 infection in school children of a rural community in central Thailand. Am J Trop Med Hyg 79:401–406

Leelayoova S, Taamasri P, Rangsin R, Naaglor T, Thathaisong U, Mungthin M (2002) In-vitro cultivation: a sensitive method for detecting *Blastocystis hominis*. Ann Trop Med Parasitol 96:803–807

Li LH, Zhang XP, Lv S, Zhang L, Yoshikawa H, Wu Z, Steinmann P, Utzinger J, Tong XM, Chen SH, Zhou XN (2007a) Cross-sectional surveys and subtype classification of human *Blastocystis* isolates from four epidemiological settings in China. Parasitol Res 102:83–90

Li LH, Zhou XN, Du ZW, Wang XZ, Wang LB, Jiang JY, Yoshikawa H, Steinmann P, Utzinger J, Wu Z, Chen JX, Chen SH, Zhang L (2007b) Molecular epidemiology of human *Blastocystis* in a village in Yunnan province, China. Parasitol Int 56:281–286

Mahmoud MS, Saleh WA (2003) Secretory and humoral antibody responses to *Blastocystis hominis* in symptomatic and asymptomatic human infections. J Egypt Soc Parasitol 33:13–30

Markell EK, Udkow MP (1986) *Blastocystis hominis*: pathogen or fellow traveler? Am J Trop Med Hyg 35:1023–1026

Mehlhorn H (1983) *Blastocystis hominis* (Brumpt 1912): Are there different stages or species? Parasitol Res 74:393–395

Mirza H, Teo JDW, Upcroft J, Tan KSW (2011a) A rapid, high-throughput viability assay for *Blastocystis spp.* reveals metronidazole resistance and extensive subtype-dependent variations in drug susceptibilities. Antimicrob Agents Chemother 55:637–648

Mirza H, Wu Z, Kidwai F, Tan KS (2011b) A metronidazole-resistant isolate of *Blastocystis* spp. is susceptible to nitric oxide and downregulates intestinal epithelial inducible nitric oxide synthase by a novel parasite survival mechanism. Infect Immun 79:5019–5026

Moe KT, Singh M, Howe J, Ho LC, Tan SW, Chen XQ, Ng GC, Yap EH (1997) Experimental *Blastocystis hominis* infection in laboratory mice. Parasitol Res 83:319–325

Nakamura Y, Hashimoto T, Yoshikawa H, Kamaishi T, Nakamura F, Okamoto K, Hasegawa M (1996) Phylogenetic position of *Blastocystis hominis* that contains cytochrome-free mitochondria, inferred from the protein phylogeny of elongation factor 1α. Mol Biochem Parasitol 77:241–245

Nigro L, Larocca L, Massarelli L, Patamia I, Minniti S, Palermo F, Cacopardo B (2003) Aplacebo-controlled treatment trial of *Blastocystis hominis* infection with metronidazole. J Travel Med 10:128–130

Nimri LF (1993) Evidence of an epidemic of *Blastocystis hominis* infections in preschool children in northern Jordan. J Clin Microbiol 31:2706–2708

Noel C, Dufernez F, Gerbod D, Edgcomb VP, Delgado-Viscogliosi P, Ho LC, Singh M, Wintjens R, Sogin ML, Capron M, Pierce R, Zenner L, Viscogliosi E (2005) Molecular phylogenies of *Blastocystis* isolates from different hosts: implications for genetic diversity, identification of species, and zoonosis. J Clin Microbiol 43:348–355

Phillips BP, Zierdt CH (1976) *Blastocystis hominis*: pathogenic potential in human patients and in gnotobiotes. Exp Parasitol 39:358–364

Silberman JD, Sogin ML, Leipe DD, Clark CG (1996) Human parasite finds taxonomic home. Nature 380:398

Stensvold CR, Suresh GK, Tan KS, Thompson RC, Traub RJ, Viscogliosi E, Yoshikawa H, Clark CG (2007) Terminology for *Blastocystis* subtypes—a consensus. Trends Parasitol 23:93–96

Tan HK, Zierdt CH (1973) Ultrastructure of *Blastocystis hominis*. Parasitol Res 42:315–324

Tan SW, Singh M, Thong KT, Ho LC, Moe KT, Chen XQ, Ng GC, Yap EH (1996) Clonal growth of *Blastocystis hominis* in soft agar with sodium thioglycollate. Parasitol Res 82:737–739

Tan TC, Ong SC, Suresh KG (2009) Genetic variability of *Blastocystis* sp. isolates obtained from cancer and HIV/AIDS patients. Parasitol Res 105:1283–1286

Wu B, Yin J, Texier C, Roussel M, Tan KSW (2010) *Blastocystis* legumain is localized on the cell surface, and specific inhibition of its activity implicates a pro-survival role for the enzyme. J Biol Chem 285:1790–1798

Yamada M, Yoshikawa H, Tegoshi T, Matsumoto Y, Yoshikawa T, Shiota T, Yoshida Y (1987) Light microscopical study of *Blastocystis* spp. in monkeys and fowls. Parasitol Res 73:527–531

Yao FR, Qiao JY, Zhao Y, Zhang X, Yang JH, Li XQ (2005) Experimental infection of mice with *Blastocystis hominis*. Zhongguo Ji Sheng Chong Xue Yu Ji Sheng Chong Bing Za Zhi 23:444–448

Yoshikawa H, Dogruman-Al F, Turk S, Kustimur S, Balaban N, Sultan N (2011) Evaluation of DNA extraction kits for molecular diagnosis of human *Blastocystis* subtypes from fecal samples. Parasitol Res 109:1045–1050

Zierdt CH, Rude WS, Bull BS (1967) Protozoan characteristics of *Blastocystis hominis*. Am J Clin Pathol 48:495–501

Zierdt CH (1973) Studies on *Blastocystis hominis*. J Protozool 20:114–121

Zierdt CH, Tan HK (1976) Ultrastructure and light microscopic appearance of *Blastocystis hominis* in a patient with enteritic disease. Parasitol Res 50:277–283

Zierdt CH, Swan JC, Hosseini J (1983) In-vitro response of *Blastocystis hominis* to antiprotozoan drugs. J Protozool 30:332–334

Chapter 2
Morphology of Human and Animal *Blastocystis* Isolates with Special Reference to Reproductive Modes

Minoru Yamada and Hisao Yoshikawa

Abstract *Blastocystis*, an anaerobic unicellular eukaryotic protozoon, is one of the most encountered microorganisms in fecal samples of various host animals including humans. The parasite is generally spherical in shape but shows a great variation in size from 5 to 50 μm. When the isolates from the feces were cultured in vitro, the size of the cell increases and some cells reach up more than 200 μm in diameter. Since a large central vacuole occupies the central part of the organisms, the cytoplasm containing nucleus, mitochondria-like organelle, and Golgi apparatus is pushed to the peripheral rim. The central vacuole possesses storage function and contains various substances such as carbohydrates, lipids, and basic proteins identified under the light and electron microscopy. These substances are accumulated in the vacuole via Golgi apparatus or crathrin-based endocytosis. There are various naming for morphologically different organisms, such as vacuolar (vacuolated), multivacuolar, avacuolar, granular, amoeboid, and cyst forms, which appear in fecal and in vitro samples. The vacuolar and granular forms are predominantly detected in the fresh feces or in vitro cultures, while some other forms, multivacuolar and avacuolar forms are rare and small in size and are mainly characterized by electron microscopy. The amoeboid (ameba) forms are also small in size and are generally more frequently observed in the samples of in vitro culture. Irregular shape of the amoeboid form is difficult to be distinguished from the amoebic forms of other intestinal protozoa, and locomotion of the amoeboid form of *Blastocystis* has not been recognized. The cyst forms are mainly seen in the fecal samples and rarely in in vitro cultures. The nuclei in vacuolar and granular forms can be clearly seen when the cells are stained with DAPI or Giemsa. In general, small cells show one or two nuclei, while bigger cells show many nuclei.

M. Yamada
Department of Infectious Diseases, Kyoto Prefectural University of Medicine, Kyoto, Japan

H. Yoshikawa (✉)
Department of Biological Sciences, Nara Women's University, Nara, Japan
e-mail: h.yoshikawa@cc.nara-wu.ac.jp

H. Mehlhorn et al. (eds.), *Blastocystis: Pathogen or Passenger?*,
Parasitology Research Monographs 4,
DOI 10.1007/978-3-642-32738-4_2,

When there were two nuclei in a cell, both the nuclei were usually located at the opposite pole of the cell. The life cycle of *Blastocystis* is still not clear. The most conflicting mode is the reproductive stages in the proposed life cycle. Binary fission is frequently observed in fresh fecal and cultured samples. However, other reproductive stages such as schizogony, plasmotomy (budding), endodyogeny, and sac-like pouches are not accepted generally for the true modes of *Blastocystis*. Schizogony-like and budding-like organisms, which possess many nuclei in the peripheral cytoplasm or have many daughter cell-like structures in the central vacuole, are observed especially in in vitro cultures. Many reproductive processes have been proposed for *Blastocystis*, however, to date, only binary fission, budding, or plasmotomy have been proven. Pseudopodal activity, on a few occasions, is seen in the amoeboid forms, and this form may be proposed as another alternative reproduction process except three identified modes of reproduction, binary fission, plasmotomy, and budding (or schizogony) under ultrastructural insights while locomotion of the amoeboid form has not still been confirmed.

2.1 Introduction

Blastocystis was first thought to be a fungus (nonpathogenic yeast), but it is now recognized as a human and animal parasitic protozoan inhabiting the cecum and large intestine (Yamada et al. 1987; Teow et al. 1992a; Boreham and Stenzel 1993; Stenzel et al. 1994; Stenzel and Boreham 1996; Abe et al. 2002; Tan 2004; Yoshikawa et al. 2004, 2007). Morphology of *Blastocystis* isolates has been extensively elucidated by light microscopy using the samples from fresh fecal materials and axenic or xenic cultures of human and animal isolates (Zierdt 1973, 1988, 1991; Yamada et al. 1987; Teow et al. 1992a; Boreham and Stenzel 1993; Zaman et al. 1993; Stenzel et al. 1994; Stenzel and Boreham 1996; Abe et al. 2002; Tan 2004; Yoshikawa et al. 2004, 2007). *Blastocystis* is strictly anaerobic eukaryotic unicellular organism, because the cells are quickly degenerated when they are exposed to the air. The most typical feature of this organism is polymorphic appearances, including a great variation in cell size and distribution of granules in the large central vacuole and polymorphic morphological shape from spherical to amoebic form. These polymorphic features are frequently observed not only within a single sample but also among isolates under conventional and phase contrast microscopy in fresh materials and in vitro cultures. Therefore, it is difficult to assign a standard morphology for diagnosis of clinical samples. Here we describe in detail, the morphological features of major forms of *Blastocystis*.

2.2 Morphology of *Blastocystis*

2.2.1 Light Microscopy

Although *Blastocystis* organisms have been isolated from a variety of animals including, mammals, birds, reptiles, amphibians, and insects, all isolates are morphologically indistinguishable (Yamada et al. 1987; Teow et al. 1992a, b; Zaman et al. 1993; Stenzel et al. 1994; Singh et al. 1996; Suresh et al. 1997; Abe et al. 2002; Yoshikawa et al. 2004). However, there are various terms for morphologically different *Blastocystis* organisms that appear in human and animal fecal samples based on the light microscopy. Namely, there are vacuolar (vacuolated), multivacuolar, avacuolar, granular, amoeboid (ameba), and cyst forms (Zierdt 1973, 1988, 1991). In general, the vacuolar and granular forms are the most predominant in the fresh fecal and in vitro culture samples. The amoeboid form is generally more frequently observed in the samples of in vitro culture than in fecal materials, while predominant excretion of the amoeboid form in stool samples of symptomatic patients has been recently reported (Tan and Suresh 2006a). Irregular shape of the amoeboid form is difficult to be distinguished from the amoebic forms of other intestinal protozoans, and locomotion of the amoeboid form of *Blastocystis* has not been recognized. The cyst form is mainly seen in the stool samples and rare in in vitro cultures. The cyst form is relatively small in size, 3–5 μm in diameter (Yoshikawa et al. 2003); hence it is difficult to identify it in the fecal samples by routine diagnostic examinations. Several cyst-concentration techniques to detect cysts from fresh fecal samples have been reported (Zaman and Khan 1994; Zaman et al. 1995).

The vacuolar and granular forms are generally round in shape and range in size from about 5–50 μm in fecal samples. When the organism is cultured, it can grow to a size of more than 50 μm and reaches up more than 200 μm in diameter. Under phase contrast and Nomarski differential interference microscopy, a large central vacuole can be clearly observed in the center of the organism. The cells having an empty vacuole are called vacuolar forms (Figs. 2.1 and 2.2). The function of the central vacuole is not yet fully confirmed. However, metabolic and storage functions of the central vacuole containing various substances such as carbohydrates and lipids have been demonstrated by light microscopy (Yoshikawa et al. 1995a, b; Yoshikawa and Hayakawa 1996a) (Fig. 2.3). Since the central vacuole occupies more than 80 % of the cell volume, the cytoplasm of the organism is pushed to the periphery and is lined thinly just under the cell membrane. During in vitro culture and also in fecal samples, some take the shape of a peanut, which indicates that the organism is in the process of cell division (binary fission) (Fig. 2.2). Many small sized or a mass of granules are sometimes observed in the central vacuole in fecal samples and cultures, and this stage is called the granular form. Granular forms in the process of binary fission can also be observed (Fig. 2.4). When *Blastocystis* organisms were stained with iodine–potassium iodide solution or Lugol's iodine, the central vacuole or containing granules can be stained yellow

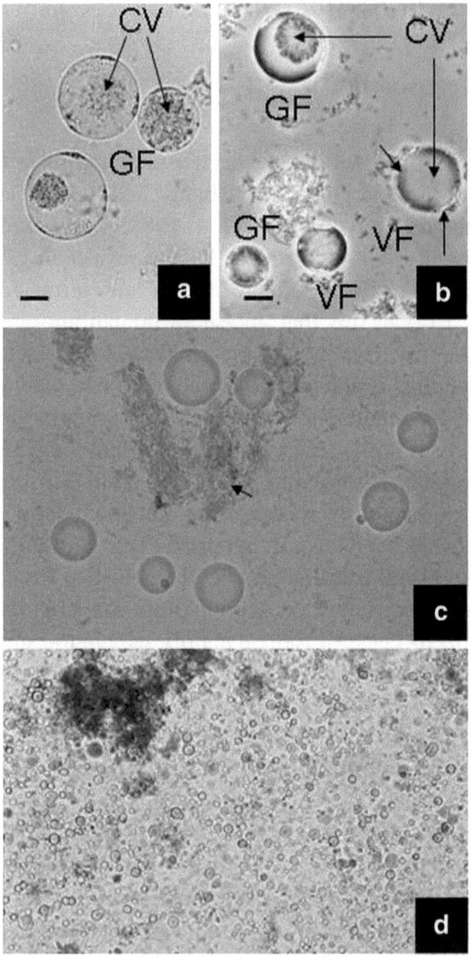

Fig. 2.1 Light microscopic images of the vacuolar and granular forms of *Blastocystis* (panels **a** and **b**). The cytoplasm (*arrows* in panel **b**) is located at the peripheral rim of the organisms due to the presence of the large central vacuole (CV). Great variations in size is evident in a culture sample and the smallest organism is observed in the center of this field (*arrow* in panel **c**, ×400). Low magnification view of the culture sample showing many morphologically different organisms (panel **d**, ×80). The conventional microscopy (panels **a**, **b**, and **c**) and phase contrast microscopy (panel **b**). Scale bar: 10 μm. *CV* central vacuole, *GF* granular form, *VF* vacuolar form

to brown, while in some cases, they are not stained at all. Staining features of Heidenhein's iron hematoxylin (HIH) are a little different to others and nonstaining areas surrounding the organisms are highlighted in the background as recognized unstained circles, due to the thick outer surface of the organisms (Fig. 2.5). This feature is also observed in the samples mixed with an India ink. Namely, circular transparent peripheral margin can be clearly seen demarcated from the surrounding

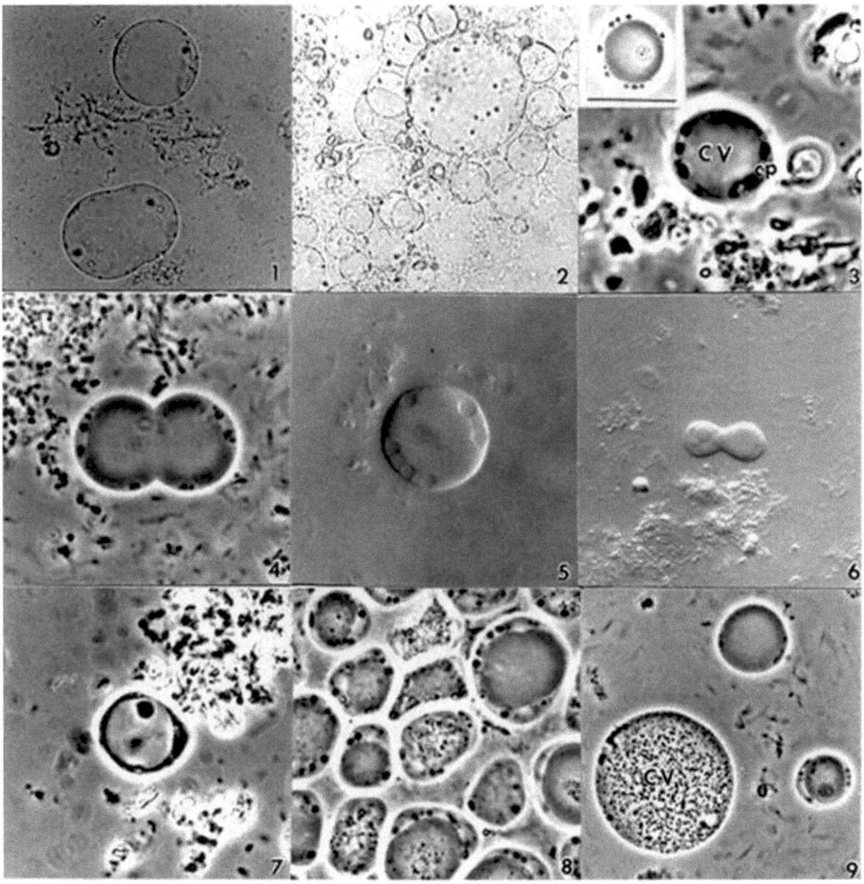

Fig. 2.2 Light microscopic images of *Blastocystis* observed conventional (panels 1 and 2), phase contrast (panels 3, 4, and 7–9), and Nomarski (differential interference) microscopy (panels 5 and 6). Panels 1 and 4, and probably 6, show organisms in progress binary fission (1, 3–5, ×800; 2, ×400; 6, ×200; 7, 8, ×600; 9, ×500)

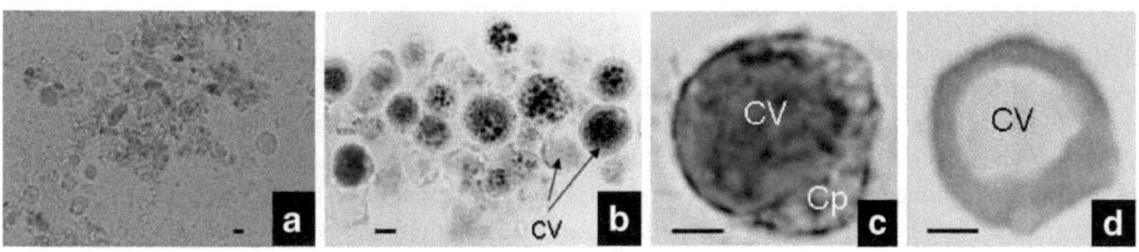

Fig. 2.3 Light micrographs of *Blastocystis* cells stained with various histochemical staining methods to detect lipids (panel **b**) and carbohydrates (panels **c** and **d**). Unstained fresh material (panel **a**) and *Blastocystis* organisms stained with Sudan black B (**b**), PAS (**c**) and Alcian blue at pH1.0 stain (**d**) Scale bars: 10 μm

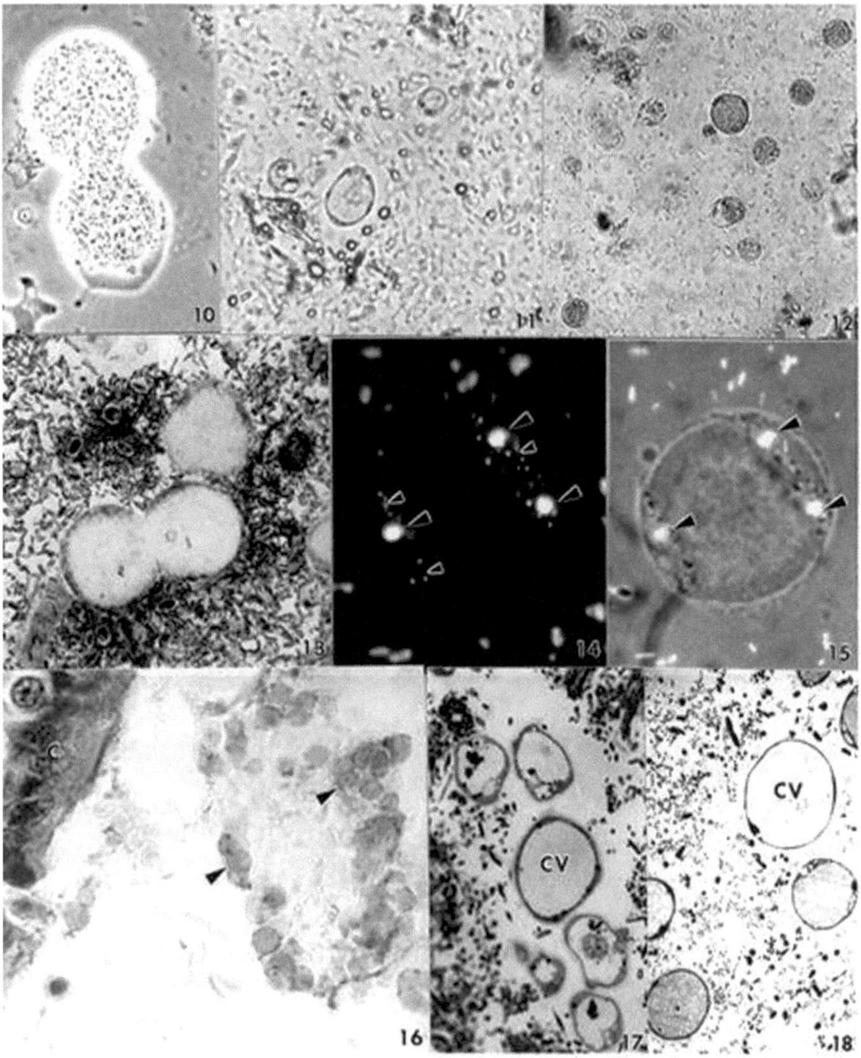

Fig. 2.4 Light microscopic appearances of *Blastocystis* organisms observed with various staining. A granular form in progress binary fission observed by the phase contrast microscopy (panel 10). Fresh materials with iodine-potassium iodide stain (Lugol's iodine) (panels 11 and 12). Note that vacuolar cells are not stained, while granular forms are well stained in the central vacuole. A diving cell stained with Giemsa (panel 13). A same organism stained with DAPI showing three nuclei (large arrowheads in panels 14 and 15) with many small dots of the DNA of mitochondria-like organelle (MLO) (small *arrowheads* in panel 14). It is evident that the thickened part of the cytoplasm contains the nucleus and MLO. A tissue section of the cecum (*c* in panel 16) stained with HE shows many *Blastocystis* in the rumen (*arrowheads* in panel 16). Semi-ultrathin sections of resin-embedded samples stained with Giemsa (panels 17 and 18) showing morphological variation of the central vacuole (panel 10, ×600; 11, 12, 16, ×200; 13, 17, 18, ×500; 14, 15, ×1,000)

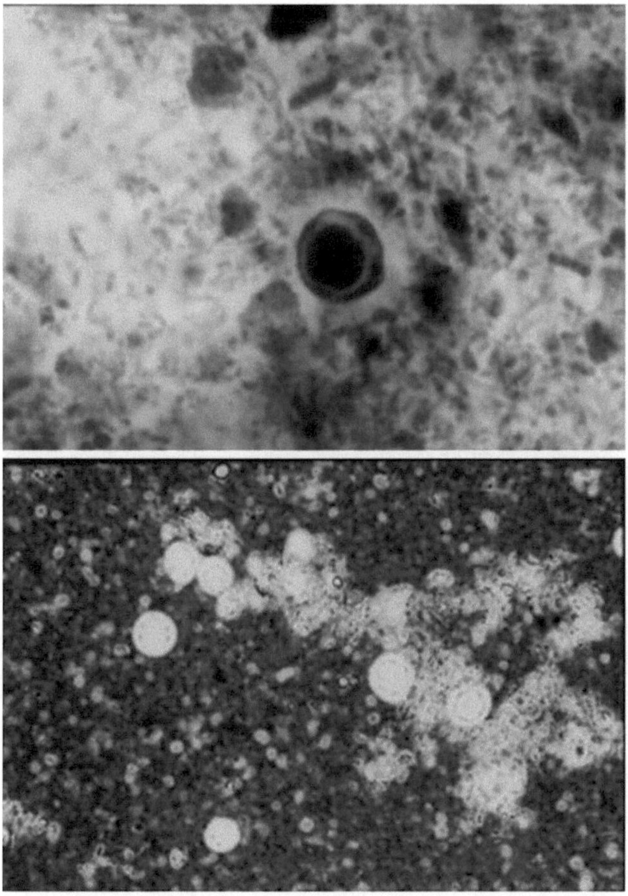

Fig. 2.5 *Blastocystis* organisms stained with Heidenhein's iron hematoxyline (HIH, *upper*) and India ink (*lower*) showing a circular transparent peripheral margin of the cells (*Upper*, ×300; *lower*, ×80)

materials (Fig. 2.5). These characteristics of *Blastocystis* organisms are hallmark for identification of the organisms correctly in fecal samples. On the other hand, organisms stained with HIH are difficult to recognize the nuclei, while the parasites stained with Giemsa can easily identify the eosinophilic nucleus and basophilic cytoplasm. This staining feature is also useful for differentiation of the parasite from other intestinal protozoan and stool debris (Fig. 2.4). *Blastocystis* isolates from monkeys usually possessed a large granule in the central vacuole under Giemsa and HIH stains (Fig. 2.6). HIH stain also shows a large granule (probably the nucleus or a volutin granule) in the peripheral cytoplasm.

In preparations stained with DAPI (4,6-diamidino-2-phenylindole), a large, well-defined, and highly fluorescent body of the nucleus is clearly observed by a fluorescence microscope (Figs. 2.4 and 2.7), while the DNA of mitochondria-like

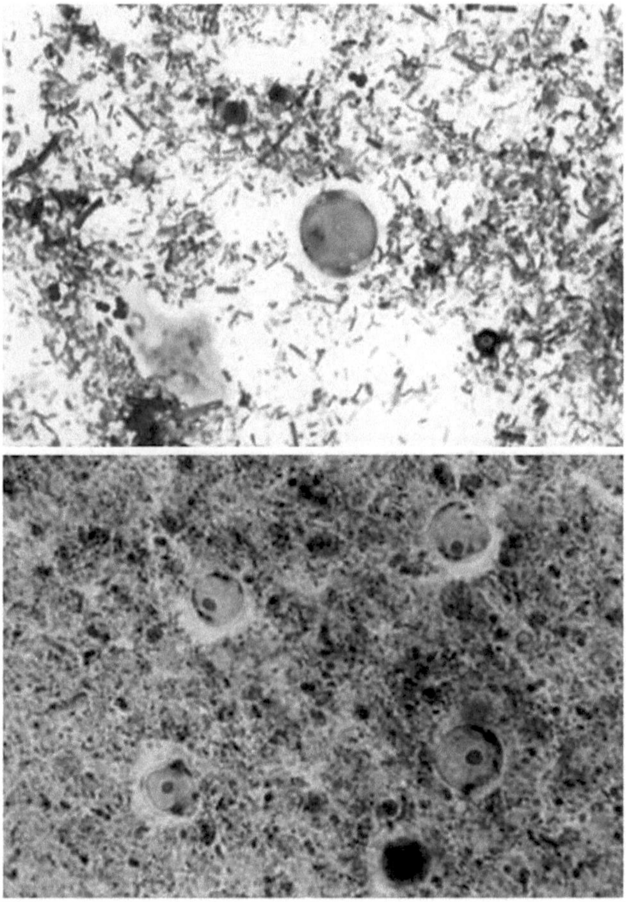

Fig. 2.6 Granular form of *Blastocystis* isolated from monkeys. Giemsa (*upper*) and HIH (*lower*) stains. A large granule can be seen in the central vacuole and unstained circular space is also observed on the outermost of the cells (*upper* ×800; *lower*, ×600)

organelle shows only small and weak fluorescence (Fig. 2.4). No fluorescence is observed within the central vacuole (Figs. 2.4 and 2.7), although schizogony and progeny in the vacuole has been reported (Zierdt 1973, 1991; Suresh et al. 1994; Singh et al. 1996).

The nucleus could also be clearly seen when fresh materials and semi-ultra thin sections are stained with Giemsa (Fig. 2.4). In general, small cells show one or two nuclei, while bigger cells show many nuclei. When there are two nuclei in a cell, both the nuclei are usually located at opposite poles of the cell (Fig. 2.1). The cytoplasm becomes thick on the side of the nucleus. The nucleus is sometimes found located in the equatorial region of the parasite.

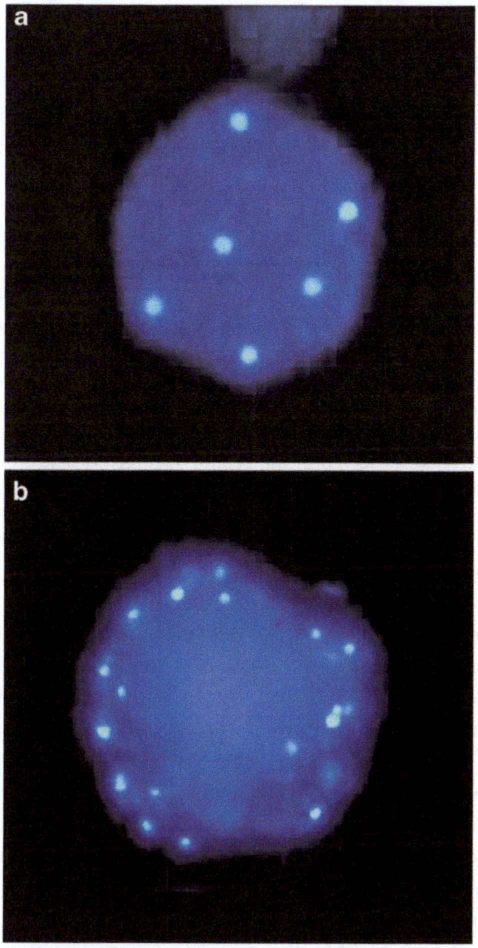

Fig. 2.7 The multi-nucleated organisms stained with DAPI observed by fluorescent light microscopy. It is evident that the nucleus of the large cell are existing on the peripheral rim of the organisms (panels **a** and **b**, ×2,000)

2.2.2 Transmission Electron Microscopy

Ultrastructure of *Blastocystis* has been extensively elucidated by transmission electron microscopy (TEM) (Zierdt 1988, 1991; Dunn et al. 1989; Stenzel et al. 1989, 1991, 1994, 1997; Stenzel and Boreham 1991, 1996; Yoshikawa and Hayakawa 1996a; Suresh et al. 1997; Pakandl 1999; Tan and Suresh 2006b; Yoshikawa et al. 2007). In general, basic morphology of *Blastocystis* organisms isolated from various host animals are very similar. However, morphological variation of the parasite is also observed within and between isolates among human origins (Dunn et al. 1989). Since a great variation of the central vacuole

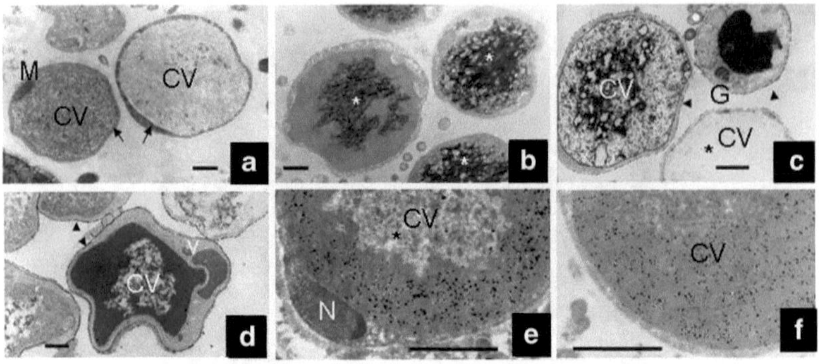

Fig. 2.8 Electron micrographs of *Blastocystis* stained with and without histochemical staining for detection of lipids, carbohydrates and basic proteins, respectively. Unstained material (panel **a**) and positive reactions with lipids (panel **b**, *asterisks*), carbohydrates (panels **c** and **d**), and basic proteins (panels **e** and **f**) are observed in the central vacuole, vesicles, or Golgi apparatus. *CV* central vacuole, *M* mitochondria-like organelle, *G* Golgi apparatus, *N* nucleus, *V* vesicles. Short *arrows* indicate the peripheral cytoplasm (panel **a**). *Arrowheads* show the filamentous layer on the surface (**c**, **d**). Each of *asterisk* in panels **c** and **f** shows no reaction with lipids and basic proteins in the central vacuoles, respectively. Scale bar: 1 μm

within a single isolate is observed in the density and distribution of the granules or materials containing the central vacuole, two distinctive vacuolar and granular forms cannot be clearly distinguished by TEM. Therefore, the terminology of vacuolar and granular forms is limited for identifying the organisms under the light microscopy. Interestingly, virus-like particles in the cytoplasm have been reported in the parasites isolated from *Macaca* monkeys (Stenzel and Boreham 1997). The presence of virus-like particles in the *Blastocystis* isolate from a seasnake, which was insensitive for DNase, while it was sensitive for RNase, was also reported (Teow et al. 1992b).

The most prominent feature of *Blastocystis* is having a large central vacuole; hence the cytoplasm is pushed in the peripheral rim. The function of the central vacuole is not yet conclusively confirmed. However, metabolic and storage functions of the central vacuole had been demonstrated by the trace of cationized ferritin added in culture medium and by the histochemical methods for lipids, carbohydrates, and basic proteins (Stenzel et al. 1989; Yoshikawa et al. 1995a, b; Yoshikawa and Hayakawa 1996a; Yoshikawa and Oishi 1997) (Fig. 2.8). Interestingly, pinocytosis via crathrin-based endocytosis on the surface of the organism clearly shows that nutrients are accumulated into the central vacuoles via small vesicles in the cytoplasm (Stenzel et al. 1989) (Fig. 2.9). It is suggested that the increasing number of the electron-dense pits on the surface membrane as the same phenomenon may indicate a change in the mode of nutrition or intensity of endocytosis (Pakandl 1999).

Blastocystis organism is surrounded by a thick outer surface coat, sometimes referred to as the fibrillar or filamentous layers or capsule. Interestingly, the thickness of the surface coat is thicker up to 1 μm among parasites freshly isolated

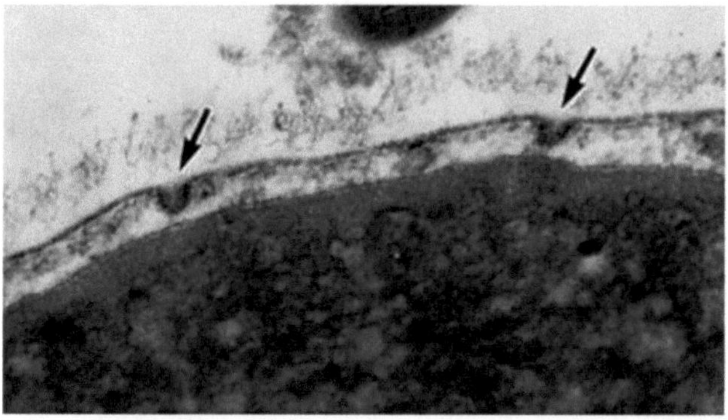

Fig. 2.9 Pinocytosis via crathrin-based endocytosis on the surface of the organism is clearly seen by electron microscopy (*arrows*) (×40,000)

from fecal samples, while it gradually becomes thin during laboratory cultures. The surface coat also contains various carbohydrates (Yoshikawa et al. 1995a; Lanuza et al. 1996). The function of the surface coat is unclear, while adherence of intestinal bacteria on the surface is frequently observed, suggesting a trapping purpose for nutrition (Dunn et al. 1989; Zaman et al. 1999).

The cytoplasm and nucleus are located on the peripheral rim of the organism due to the presence of the large central vacuole. The cytoplasm of anaerobic *Blastocystis* organisms contains many organelles commonly observed in eukaryotic cells, namely, various-sized vacuoles, mitochondria-like organelle (MLO), Golgi apparatus, rough endoplasmic reticulum, microtubules, and ribosomes in the cytoplasm (Fig. 2.10). Binary fission of the vacuolar/granular form is frequently observed, while this division is rare in the multivacuolar cells. However, nuclear division with mitotic apparatus does not yet observed. Interestingly, two daughter nuclei are frequently seen separated by a divided inner membrane, the whole still enclosed within an intact outer membrane (Fig. 2.11). Same observation was also reported in *Blastocystis* isolates from pigs (Pakandl 1999) and by freeze-fracture electron microscopy of human isolates (Yoshikawa et al. 1988). This observation lends credence to the theory that nuclear division of *Blastocystis* begins with separation of the parent nucleoplasm by the inner membrane into two parts, followed by division of the outer membrane forming two discrete nuclei. This mode of nuclear division is reported to occur in the mitosis of *Eimeria necatrix* (Dubremetz 1973).

Blastocystis is strictly anaerobic organism, and it possesses many electron-dense organelles which morphologically resemble mitochondria (Figs. 2.10 and 2.11). Since it is known to lack cytochrome C complex genes (Wawrzyniak et al. 2008), this organelle is now called as MLO. The function of the MLO is not known, while it may play a similar role in energy metabolism as classic mitochondria (Hamblin et al. 2008). The MLO of *Blastocystis* are either spherical or oval in shape, 1–3 μm in diameter, and showed many tubular or sacculate cristae which are typical feature of eukaryotic protozoa. Interestingly, under high magnification, spiral structures consisting of intertwined coils are observed inside each tubular crista (Fig. 2.12).

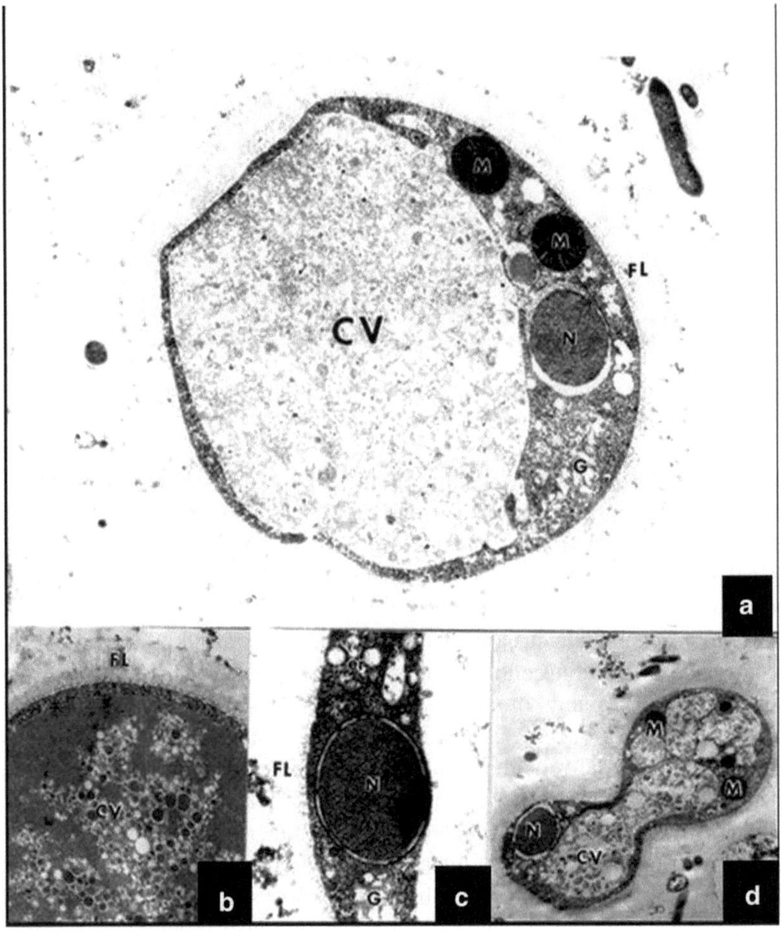

Fig. 2.10 Ultrastructure of *Blastocystis* organisms. The central vacuole (CV) showed various density and distribution of granules among the organisms (panels **a–d**). High magnification view of the cytoplasm showing a nucleus (*N*), Golgi apparatus (*G*), and endoplasmic reticulum (*er*) (panel **c**). A granular form in progress binary fission (panel **d**). *CV* central vacuole, *FL* filamentous layer, *G* Golgi apparatus, *M* mitochondria-like organelle, *N* nucleus (panel **a** ×1,000, **b** × 20,000, **c** ×13,000, **d** × 5,000)

The Golgi apparatus of *Blastocystis* is always found adjacent to the nucleus (Figs. 2.8 and 2.10). Small electron-dense secretory granules are often seen near the Golgi apparatus. As general function of this organelle, synthesis of carbohydrate has been demonstrated histochemically at ultrastructural level (Yoshikawa et al. 1995a) (Fig. 2.8).

Since the cyst form is relatively small in size, it is difficult to identify it by light microscopy; hence the cyst form has been examined by electron microscopy in human and animal isolates (Stenzel and Boreham 1991; Zaman et al. 1993, 1995; Stenzel et al. 1997; Chen et al. 1999; Moe et al. 1996; Yoshikawa et al. 2003).

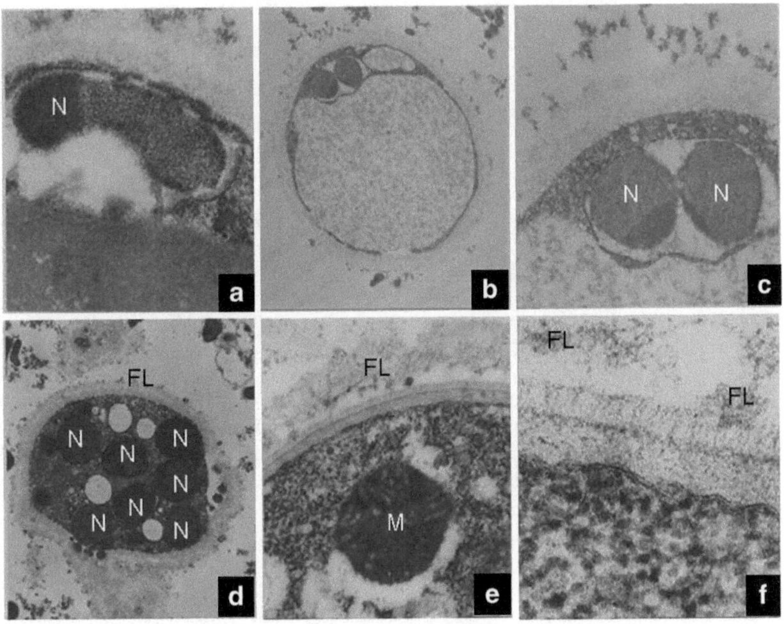

Fig. 2.11 Nuclear division and cyst form of *Blastocystis* of human isolates in cultures and fecal samples, respectively. Dividing nucleus or two daughter nuclei surrounding the outer nuclear envelope was frequently observed in the cultured samples (panels **a–c**). It is evident of the thick surface filamentous layer of the organisms after the short in vitro cultures. The multi-nucleate cell shows seven nuclei in a section of thickened cytoplasm (panel **d**). A cyst form observed in human fecal samples showed a three-layered thick cyst wall surrounding with filamentous materials on the outer surface (panels **e** and **f**) which may be corresponding to the outer filamentous layer existing on the surface of the vacuolar and granular forms (panels **a–d**). *FL* filamentous layer, *M* mitochondria-like organelle, *N* nucleus (panels **a**, **c**, **e** × 20,000; **b** × 5,000; **d** × 6,500; **f** × 95,000)

The morphological characteristics of the cyst form reveal a thick wall and a constant average size in diameter, 3–5 μm, except for the larger size of an isolate from a monkey (Stenzel et al. 1997) (Fig. 2.12). Although the size of the cyst form is relatively small, it contains one to 4 nuclei (Yoshikawa et al. 2003). Since a thick cyst wall is composed beneath the outermost surface coat, the outermost surface coat (filamentous layer) seems to be fuzzy (Figs. 2.11 and 2.14). In general, the cyst form is only detected in the fecal samples of human and animals, and rarely in in vitro cultures (Stenzel and Boreham 1991; Zaman et al. 1993, 1995; Stenzel et al. 1997; Chen et al. 1999; Moe et al. 1996; Yoshikawa et al. 2003). Interestingly, the central vacuole does not exist in this form, while a large mass of the glycogen granules in the cytoplasm is reported in a reptilian isolate (Yoshikawa et al. 2003) (Fig. 2.13). The number of nucleus is also variable among the isolates and it has been reported from a single nucleus to 4 nuclei (Stenzel and Boreham 1991; Zaman et al. 1993, 1995; Stenzel et al. 1997; Chen et al. 1999; Yoshikawa et al. 2003).

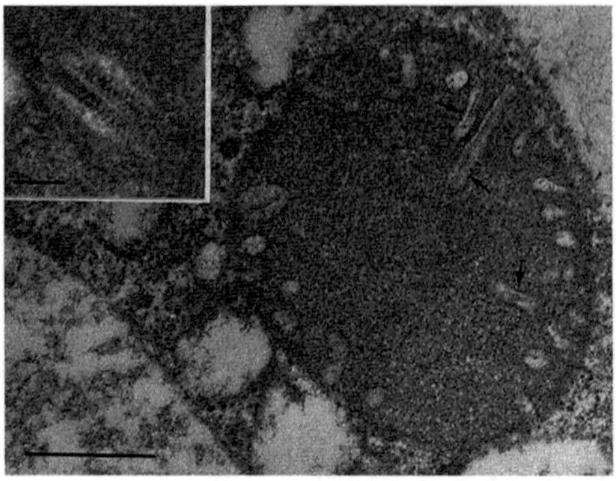

Fig. 2.12 High magnification view of the mitochondria-like organelle (MLO). It is evident a spiral structures consisted of intertwined coils in the rumen of each tubular crista of the MLO. Scale bars: 0.5 μm (**a**), 0.1 μm (**b**). This figure was cited from Fig. 2.13 in page 1490 in Yoshikawa T (1988) Fine structure of *Blastocystis hominis* Brumpt, 1912. J Kyoto Pref Univ Med 97:1483–1500

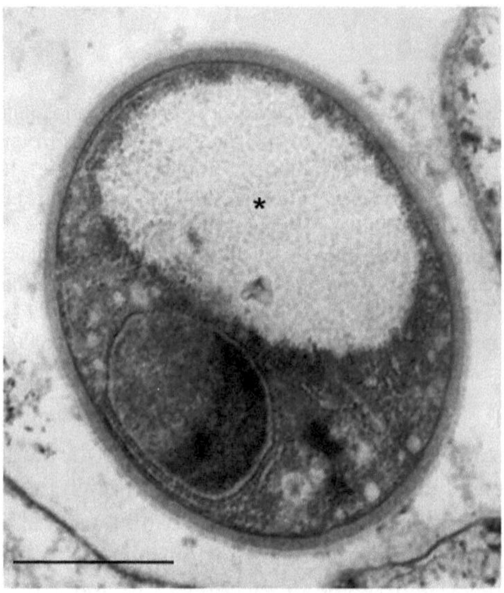

Fig. 2.13 A typical cyst form of *Blastocystis* reptilian isolate having a mass of glycogen granule (*asterisk*) Scale bar: 1 μm

2.2.3 Freeze-Fracture Electron Microscopy

Ultrastructure of a human *Blastocystis* isolate has also been elucidated by freeze-fracture electron microscopy (Yoshikawa et al. 1988; Yoshikawa and Hayakawa 1996b). The most remarkable feature of *Blastocystis* is the heterogeneous

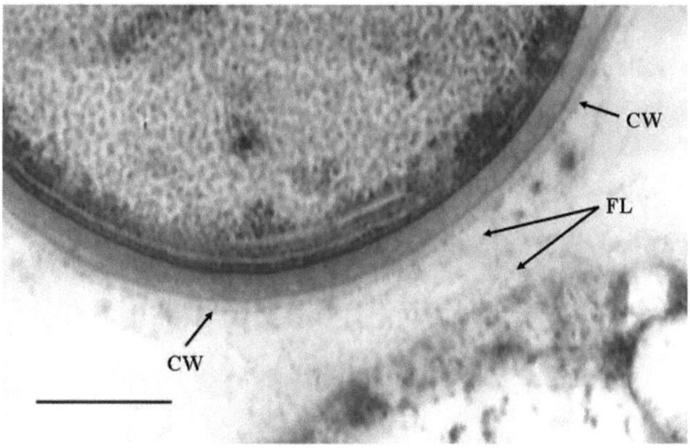

Fig. 2.14 High magnification view of the cyst form and vacuolar form of *Blastocystis* reptilian isolate. It is evident that cyst wall is composed of three layers, while the vacuolar form shows only filamentous outer layer on the surface. Scale bar: 0.5 μm

distribution of the intramembranous particles (IMPs) of the plasma membrane because IMPs, known to be integral membrane proteins, are generally distributed homogeneously in the plasma membrane (Yoshikawa et al. 1988). Similarly, membrane cholesterol is also distributed heterogeneously associated with membrane proteins (Yoshikawa and Hayakawa 1996b). In addition, different organization of the membrane proteins and cholesterol between the plasma membrane and central vacuole membrane demonstrate different organization of the two membranes (Yoshikawa et al. 1988, Yoshikawa and Hayakawa 1996b). A great variation of the distribution and density of the granules in the central vacuole was also observed. As TEM images, two daughter nuclei enclosed within an intact outer membrane are also clearly observed by freeze-fracture electron microscopy (Yoshikawa et al. 1988). When the cyst form of reptilian isolate of in vitro culture was observed by freeze-fracture methods, the protoplasmic face of the plasma membrane of cyst form showed extremely low density of the IMPs with many striated indentations (Fig. 2.15), while non-cyst form (i.e. vacuolar and granular form) showed many IMPs (Yoshikawa et al. 2003). Practically no IMPs in cyst form support the resting stage of this form because most of the IMPs have enzymatic activity or transport function of the membrane. The fractured cyst wall surrounding the small-sized cyst form show many fine granules (Fig. 2.15), suggesting protein constituents within filamentous materials observed by TEM (Fig. 2.11).

2.2.4 Scanning Electron Microscopy

A great variation in cell size among a human *Blastocystis* isolate and smooth surface and round shape of the organisms are also evident by scanning electron

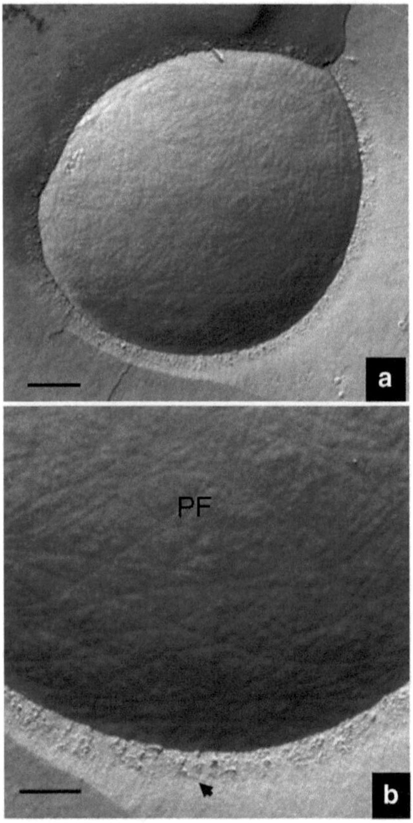

Fig. 2.15 Freeze-fracture images of *Blastocystis* cyst form from reptilian isolate in in vitro culture. The P-face (PF) of the plasma membrane of the cyst form of a reptilian isolate showed an extremely low density of IMPs and many striated indentations. Many fine granules are seen on the fractured cyst wall (*arrow*). Scale bar in panel **a** 1 μm, Scale bar in panel **b** 0.2 μm

microscopy (SEM) (Fig. 2.16) (Matsumoto et al. 1987). Most of the organisms in in vitro cultures show spherical shape, while binary fission is also observed (Fig. 2.16). The cell surface of the organisms from humans, monkeys, pigs, and chicken fecal samples has been compared with a cultured human isolate (Cassidy et al. 1994). It is apparent that the surface structures of *Blastocystis* isolates from different hosts are variable, and the organisms from culture are not typical of the fresh fecal samples. Namely, a cultured sample showed much thinner surface coat and smooth outer surface than those of the fecal samples. However, these morphological differences among the isolates are insufficient to differentiate speciation (Cassidy et al. 1994). Interestingly, *Blastocystis* organisms in a chicken fecal sample show surface coat projections. Similarly, fibrillar structure of the surface coat and individual fibrils extending up to 5 μm from the periphery of the organism are also observed in human *Blastocystis* isolates (Zaman et al. 1999). In some cells,

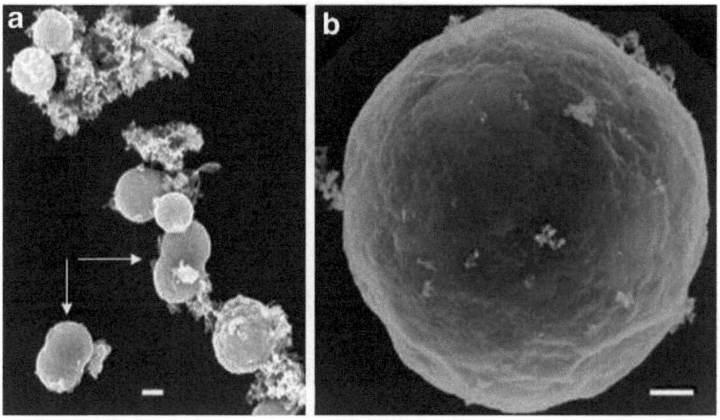

Fig. 2.16 *Blastocystis* organisms observed by scanning electron microscopy (SEM). Note the cell division (binary fission) (*arrows*) among spherical organisms in in vitro culture (**a**). The surface of the organism is uneven (**b**). Scale bars: 10 μm (**a**), 5 μm (**b**)

bacteria adhering to the surface are also observed. Although the function of the surface coat is not known, entrapment mechanism for bacteria for nutritive purpose and adherence to the epithelial lining of the gut are speculated (Zaman et al. 1999).

2.2.5 Reproductive Modes

2.2.6 Binary Fission

The life cycle of *Blastocystis* is not yet conclusively demonstrated. Many hypotheses for reproduction mode of *Blastocystis* have been put forward, though little supportive evidence exists (Suresh et al. 1994, 1997; Govind et al. 2002, 2003; Tan and Stenzel 2003; Windsor et al. 2003). The current only accepted mode of reproduction for *Blastocystis* is binary fission (Matsumoto et al. 1987; Dunn et al. 1989; Stenzel and Boreham 1996; Moe et al. 1996; Govind et al. 2002; Tan 2008). When DNA content of a single nucleus among various organisms in in vitro culture of human *Blastocystis* isolate was randomly measured by DAPI staining with fluorescence microscopy using the same method as reported in *Pneumocystis* organism (Yamada et al. 1986), two peaks corresponding to 1C and 2C was observed (Fig. 2.17). In addition, when DNA content of the each single nucleus was compared between a single- and two-nucleated cells, the lower or higher DNA content nuclei than 1C cells was observed among the two-nucleated organisms, respectively (Fig. 2.17). These results strongly support that some cells are in progress of nuclear division. Therefore, binary fission truly exists in the life cycle of *Blastocystis*.

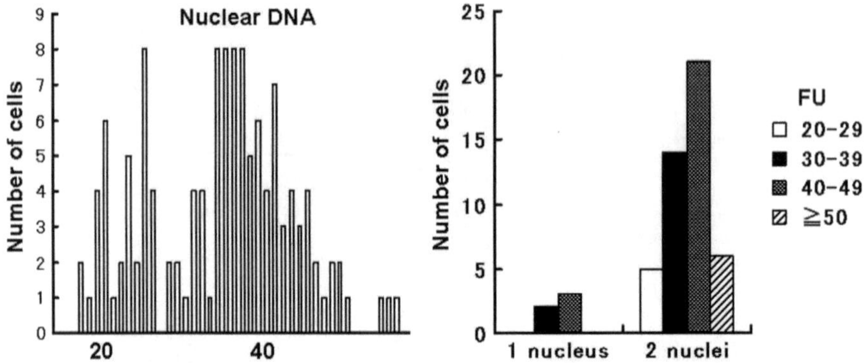

Fig. 2.17 DNA contents of *Blastocystis* nucleus measured by fluorescence microscopy with DAPI staining. *Left* Distribution of a single nuclear fluorescence intensity among *Blastocystis* organisms containing various number of the nucleus. It is evident of two peaks of the DNA contents showing nuclear division. Abscissa: fluorescence intensity in arbitrary unit. Ordinate: number of organisms. *Right* Distribution of a single nuclear fluorescence intensity among *Blastocystis* organisms containing a single nucleated cells and two nuclei cells. Most of the DNA contents of a each nucleus are equal among the both *Blastocystis* organisms, while a few number of the nucleus of the organisms having two nuclei show a lower or higher DNA contents than the two peaks of the organisms having a single nuclei. Abscissa: organisms containing a nucleus or two nuclei. *Ordinate* number of organisms. *FU* fluorescence intensity in arbitrary unit

2.2.7 Other Reproductive Modes

The additional reproductive modes of *Blastocystis* have also been reported. These are plasmotomy, budding, schizogony, endodyogeny, endogeny, and sac-like pouches (indicating the production of progeny) (Zierdt 1988; Suresh et al. 1997; Govind et al. 2002; Windsor et al. 2003; Tan and Suresh 2007). However, these reproductive modes are not generally accepted as truly present in the life cycle of *Blastocystis*. The schizogony (progeny reproduction) has been reasoned under phase-contrast microscopy by Zierdt (1991) and multiple fission or schizogony (plasmotomy-like) has been also characterized (Zaman 1997), but under the level of light microscopy, it is difficult to determine this mode clearly. Tan and Stenzel (2003) suggested that such progeny as a model of reproduction must be shown to contain genetic material and to develop into viable adult cells. In any case, care should be taken when describing reproductive processes in *Blastocystis* using light microscopy alone and these modes should be necessary to support electron microscopy (Windsor et al. 2003), while in in vitro cultures five modes of reproduction, namely, binary fission, endodyogeny, plasmotomy, budding, and schizogony are frequently observed (Zhang et al. 2007) (Figs. 2.18, 2.19, 2.20, 2.21, 2.22, 2.23, and 2.24). Tan and Suresh (2007) also confirmed that plasmotomy, leading to the formation of multiple nuclei, was the other reproductive mode of *Blastocystis* other than binary fission. Under Giemsa and HIH stains, fresh samples of the feces and in vitro cultures frequently show polymorphic budding organisms

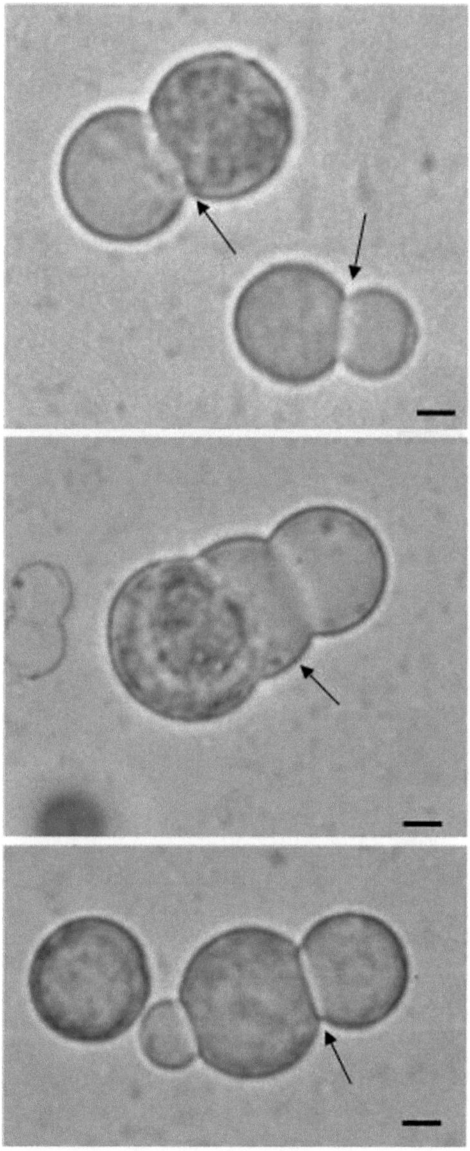

Fig. 2.18 Various reproductive modes of *Blastocystis* organisms (Figs. 2.18–2.23). A typical binary fission and budding form of vacuolar and granular forms. In vitro cultures, two or three dividing or budding organisms are frequently seen (*arrows*). Scale bar: 10 μm

(three-dividing organisms) and schizogony-like organisms showing multi-nuclei in the central part of the organisms (Figs. 2.18, 2.19, 2.20). Scanning electron microscopy showed that the cell division with binary fission is the main reproductive stage but budding-like organisms are rarely observed (Fig. 2.22). When observing cells

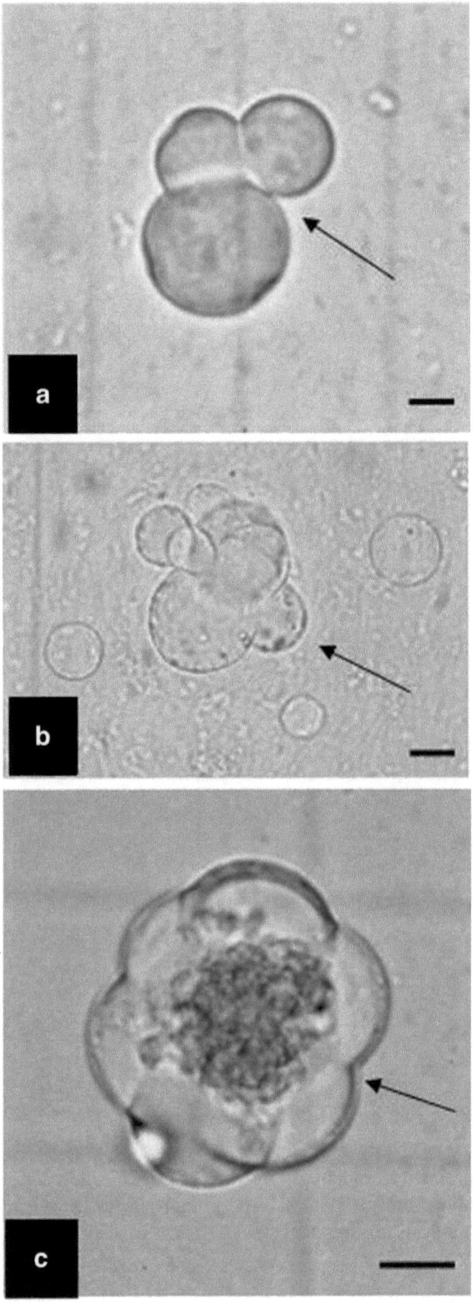

Fig. 2.19 An amoeba-like or budding-like organism (panel **a**), multi-budding (panel **b**), and schizogony-like (panel **c**) organisms are observed in vitro cultures (*arrows*). Scale bar: 10 μm

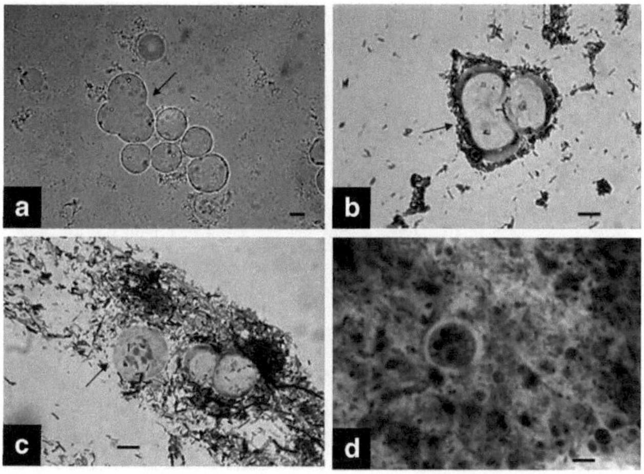

Fig. 2.20 Triple-budding-like organisms (*arrows*) observed in fresh specimen (panel **a**) and Giemsa stain (panel **b**) in culture samples. Schizogony-like organisms (*arrows*) stained with Giemsa (panel **c**) and HIH (panel **d**) in fresh human fecal samples. It is evident of multi-nuclei in the central part of the organisms (panels **c** and **d**). Scale bar: 10 μm

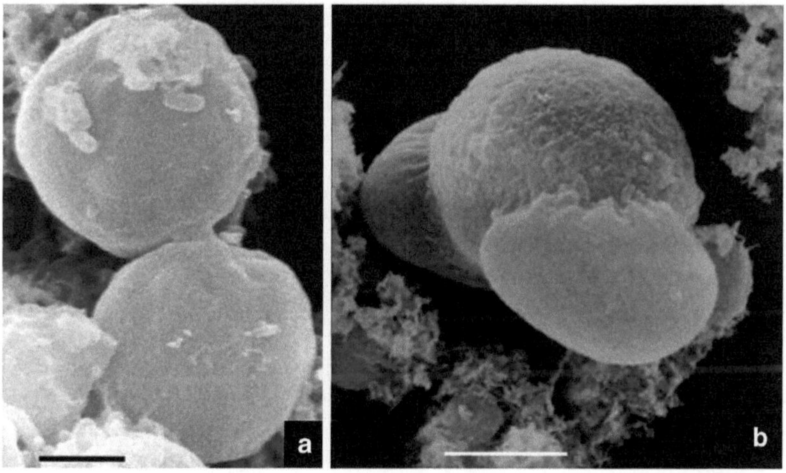

Fig. 2.21 *Blastocystis* organisms observed by scanning electron microscopy (SEM). The surface of the organism is uneven (panel **a**) and a budding-like organism is also observed (panel **b**). Scale bars: 5 μm (**a**), 10 μm (**b**)

in vitro cultures, schizogony-like organisms, which have many daughter cell-like structures in the central vacuole, are found (Fig. 2.22). Schizogony-like organisms, which possess many daughter cells in the central vacuole, are rarely seen in human fecal samples (Fig. 2.21c, d). In addition, under in vitro culture, budding-like organisms sometimes show many nuclei in each dividing cytoplasm (Fig. 2.24).

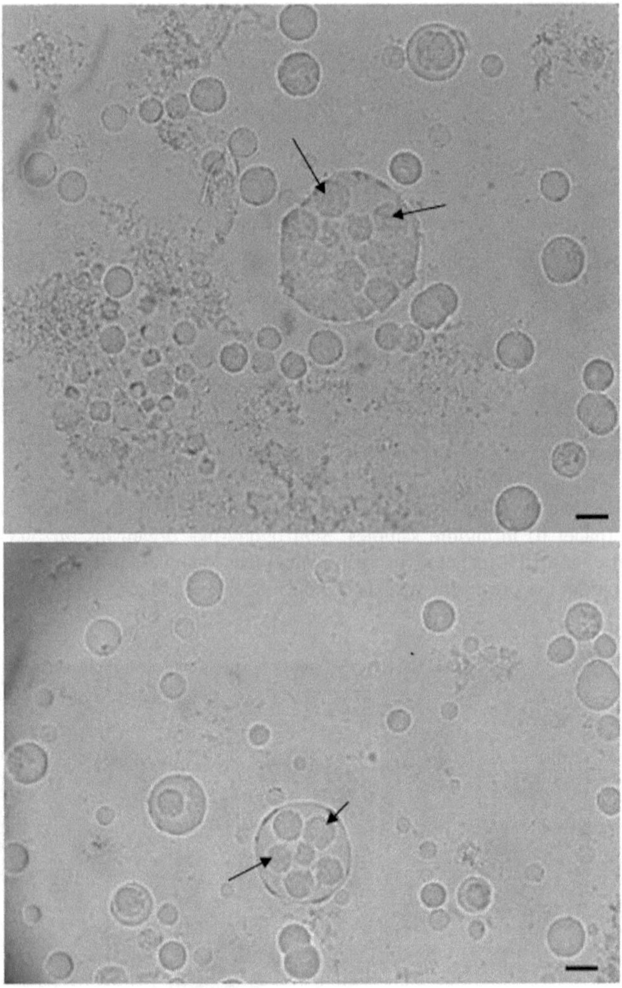

Fig. 2.22 Schizogony-like organisms observed in culture samples. Several daughter cell-like structures are seen in the central vacuole (*arrows*). Scale bar: 10 μm

Interestingly, however, these endodyogeny, plasmotomy, budding, and schizogony have not yet recognized by TEM.

The amoeboid forms of *Blastocystis* have been observed on a few occasions as the other alternative reproduction process (Tan and Suresh 2007) and they may indicate pseudopodal activity (Fig. 2.24). However, irregular shape is more likely artifact in the processing for TEM (Tan 2004) and disruption of this anaerobic parasite with exposing air is also concerned because locomotion of amoeboid form has not been yet observed.

Many reproductive processes have been suggested for *Blastocystis*, however, to date, only binary fission, budding, or plasmotomy have been proven. It is true that

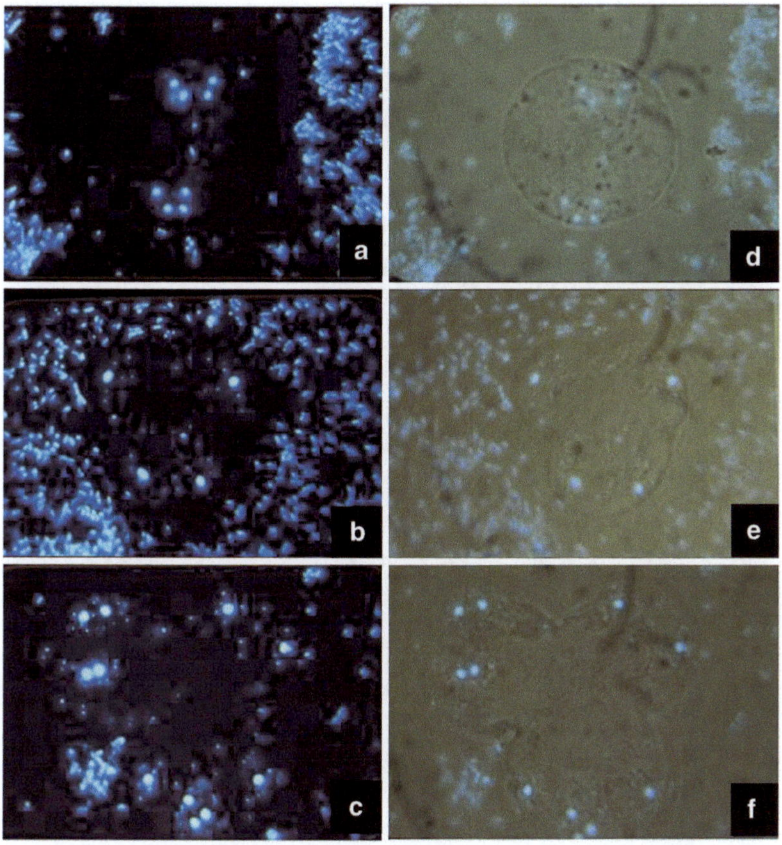

Fig. 2.23 Schizogony-like (panels **a** and **d**) and budding-like organisms (panels **b**, **c** and **e**, **f**) stained with DAPI observed by fluorescence microscopy (*left*, panels **a–c**) and conventional microscopy (*right*, panels **d–f**), respectively. It is evident that nuclei are localized in the central part of the cell showing a schizogony-like cell, while the several nuclei are separated into the budding cells (×2,000)

the reproductive modes of *Blastocystis* are binary fission, budding, and possibly schizogony. The mode of schizogony, supportive of the production of progeny, has been only confirmed with light microscope observations and has not been clearly determined with electron microscopy. The amoeboid form may be proposed as another alternative reproduction process (Fig. 2.22). Recently, three modes of reproduction, binary fission, plasmotomy, and budding were proposed under ultrastructural insights (Zhang et al. 2011). Since the available data for identifying the other reproductive modes, plasmotomy, schizogony, and division of amoeboid form are only limited at present, further detail research on the other reproductive modes will be required to clarify whether these modes are truly exist on the life cycle of *Blastocystis*.

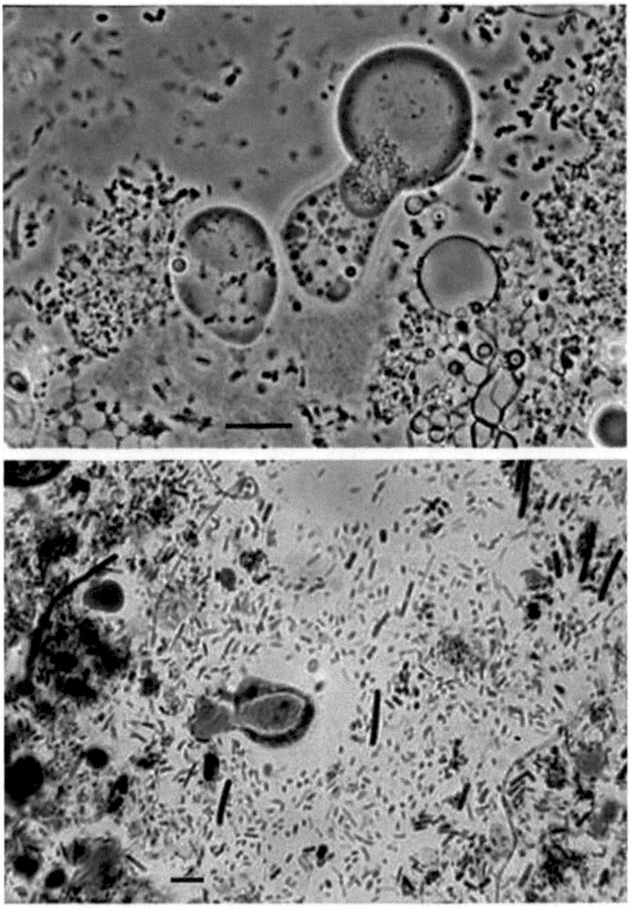

Fig. 2.24 The other alternative reproduction process of *Blastocystis* observed by phase-contrast microcopy and HIH stain. These morphologically irregular cells may show pseudopodal activity. Scale bar: 10 μm

2.3 Conclusions

Although *Blastocystis* is one of the common intestinal protozoa found in human and animal fecal samples, the taxonomy and reproduction modes of the organism are still not fully understood. Some of morphological appearances in light and electron microscopy are controversial surrounding this parasite is the lack of the uniformed consents. However, the recent accumulating evidences on the morphology and reproductive modes of various isolates promote that *Blastocystis* is more mysterious eukaryote and contribute to its biology gradually but steadily. Researches on the morphological approaches and identifications of the true reproductive stages or modes in the life cycle should be useful for the interpretation of a parasite infecting in the broad range of hosts from vertebrate to invertebrate animals.

References

Abe N, Nagoshi M, Takami K, Sawano Y, Yoshikawa H (2002) A survey of *Blastocystis* sp. in livestock, pets, and zoo animals in Japan. Vet Parasitol 106:203–212

Boreham PFL, Stenzel DJ (1993) *Blastocystis* in humans and animals: morphology, biology, and epizootiology. Adv Parasitol 32:1–70

Cassidy MF, Stenzel DJ, Boreham PFL (1994) Electron microscopy of surface structures of *Blastocystis* sp. from different hosts. Parasitol Res 80:505–511

Chen XQ, Singh M, Howe J, Ho LC, Tan SW, Yap EH (1999) In vitro encystation and excystation of *Blastocystis ratti*. Parasitology 118:151–160

Dubremetz JF (1973) Étude ultrastructurale de la mitose schizogonique chez la coccidie *Eimeria necatrix* (Johnson 1930). J Ultrastruct Res 42:354–376

Dunn LA, Boreham PFL, Stenzel DJ (1989) Ultrastructural variation of *Blastocystis hominis* stocks in culture. Int J Parasitol 19:43–56

Govind SK, Khairul AA, Smith HV (2002) Multiple reproductive processes in *Blastocystis*. Trends Parasitol 18:528

Govind SK, Anuar KA, Smith HV (2003) Response to Tan and Stenzel, and Windsor *et al.*: *Blastocystis* reproduction and morphology. Trends Parasitol 19:291–292

Hamblin K, Standley DM, Rogers MB, Stechmann A, Roger AJ, Maytum R, van der Giezen M (2008) Localization and nucleotide specificity of *Blastocystis* succinyl-CoA synthetase. Mol Microbiol 68:1395–1405

Lanuza MD, Carbajal JA, Borrás R (1996) Identification of surface coat carbohydrates in *Blastocystis hominis* by lectin probes. Int J Parasitol 26:527–532

Matsumoto Y, Yamada M, Yoshida Y (1987) Light-microscopical appearance and ultrastructure of *Blastocystis hominis*, an intestinal parasite of man. Zbl Bakt Microbiol Hyg Ser A 264:379–385

Moe KT, Singh M, Howe J, Ho LC, Tan SW, Ng GC, Chen XQ, Yap EH (1996) Observations on the ultrastructure and viability of the cystic stage of *Blastocystis hominis* from human feces. Parasitol Res 82:439–444

Pakandl M (1999) *Blastocystis* sp. from pigs: ultrastructural changes occurring during polyxenic cultivation in Iscove's modified Dulbecco's medium. Parasitol Res 85:743–748

Singh M, Ho LC, Yap ALL, Ng GC, Tan SW, Moe KT, Yap EH (1996) Axenic culture of reptilian *Blastocystis* isolates in monophasic medium and speciation by karyotypic typing. Parasitol Res 82:165–169

Stenzel DJ, Dunn LA, Boreham PFL (1989) Endocytosis in cultures of *Blastocystis hominis*. Int J Parasitol 19:787–791

Stenzel DJ, Boreham PFL (1991) A cyst-like stage of *Blastocystis hominis*. Int J Parasitol 21:613–615

Stenzel DJ, Boreham PFL, McDougall R (1991) Ultrastructure of *Blastocystis hominis* in human stool samples. Int J Parasitol 21:807–812

Stenzel DJ, Cassidy MF, Boreham PFL (1994) Morphology of *Blastocystis* sp. from domestic birds. Parasitol Res 80:131–137

Stenzel DJ, Boreham PFL (1996) *Blastocystis hominis* revisited. Clin Microbiol Rev 9:563–584

Stenzel DJ, Boreham PFL (1997) Virus-like particles in *Blastocystis* sp. from simian faecal materials. Int J Parasitol 27:345–348

Stenzel DJ, Lee MG, Boreham PFL (1997) Morphological differences in *Blastocystis* cysts—an indication of different species? Parasitol Res 83:452–457

Suresh K, Howe J, Ng GC, Ho LC, Ramachandran NP, Loh AK, Yap EH, Singh M (1994) A multiple fission-like mode of asexual reproduction in *Blastocystis hominis*. Parasitol Res 80:523–527

Suresh K, Mak JW, Chuong LS, Ragunathan T, Init I (1997) Sac-like pouches in *Blastocystis* from the house lizard *Cosymbotus platyurus*. Parasitol Res 83:523–525

Tan KSW (2004) *Blastocystis* in humans and animals: new insights using modern methodologies. Vet Parasitol 126:121–144

Tan KSW, Stenzel DJ (2003) Multiple reproductive processes in *Blastocystis*: proceed with caution. Trends Parasitol 19:290–291

Tan TC, Suresh KG (2006a) Predominance of amoeboid forms of *Blastocystis hominis* in isolates from symptomatic patients. Parasitol Res 98:189–193

Tan TC, Suresh KG (2006b) Amoeboid form of *Blastocystis hominis*—a detailed ultrastructural insight. Parasitol Res 99:737–742

Tan TC, Suresh KG (2007) Evidence of plasmotomy in *Blastocystis hominis*. Parasitol Res 101:1521–1525

Tan KSW (2008) New insights on classification, identification, and clinical relevance of *Blastocystis* spp. Clin Microbiol Rev 21:639–665

Teow WL, Ng GC, Chan PP, Chan YC, Yap EH, Zaman V, Singh M (1992a) A survey of *Blastocystis* in reptiles. Parasitol Res 78:453–455

Teow WL, Ho LC, Ng GC, Chan YC, Yap EH, Chan PP, Howe J, Zaman V, Singh M (1992b) Virus-like particles in a *Blastocystis* species from the seasnake, *Lapemis hardwickii*. Int J Parasitol 22:1029–1032

Wawrzyniak I, Roussel M, Diogon M, Couloux A, Texier C, Tan KSW, Vivares CP, Delbac F, Wincker P, EL Alaoui H (2008) Complete circular DNA in the mitochondria-like organelles of *Blastocystis hominis*. Int J Parasit 38:1377–1382

Windsor JJ, Stenzel DJ, Macfarlane L (2003) Multiple reproductive processes in *Blastocystis hominis*. Trends Parasitol 19:289–290

Yamada M, Matsumoto Y, Hamada S, Fujita S, Yoshida Y (1986) Demonstration and determination of DNA in *Pneumocystis carinii* by fluorescence microscopy with 4′,6-diamidino-2-phenylindole (DAPI). Zbl Bakt Microbiol Hyg Ser A 262:240–246

Yamada M, Yoshikawa H, Tegoshi T, Matsumoto Y, Yoshikawa T, Shiota T, Yoshida Y (1987) Light microscopical study of *Blastocystis* spp. in monkeys and fowls. Parasitol Res 73:527–531

Yoshikawa H, Hayakawa A (1996a) Morphological changes in the central vacuole of *Blastocystis hominis* during in vitro culture. Protoplasma 194:63–68

Yoshikawa H, Hayakawa A (1996b) Freeze-fracture cytochemistry of membrane cholesterol in *Blastocystis hominis*. Int J Parasitol 26:1111–1114

Yoshikawa H, Oishi K (1997) Ultrastructural localization of basic proteins of *Blastocystis hominis*. Protoplasma 200:31–34

Yoshikawa H, Yamada M, Yoshida Y (1988) Freeze-fracture study of *Blastocystis hominis*. J Protozool 35:522–528

Yoshikawa H, Kuwayama N, Enose Y (1995a) Histochemical detection of carbohydrates of *Blastocystis hominis*. J Eukaryot Microbiol 42:70–74

Yoshikawa H, Satoh J, Enose Y (1995b) Light and electron microscopic localization of lipids in *Blastocystis hominis*. J Electron Microsc 44:100–103

Yoshikawa H, Nagashima M, Morimoto K, Yamanouti Y, Yap EH, Singh M (2003) Freeze-fracture and cytochemical studies on the in vitro cyst form of reptilian *Blastocystis pythoni*. J Eukaryot Microbiol 50:70–75

Yoshikawa H, Morimoto K, Nagashima M, Miyamoto N (2004) A survey of *Blastocystis* infection in anuran and urodele amphibians. Vet Parasitol 122:91–102

Yoshikawa H, Wu Z, Howe J, Hashimoto T, Geok-Choo NG, Tan KSW (2007) Ultrastructural and phylogenetic studies on *Blastocystis* isolates from cockroaches. J Eukaryot Microbiol 54:33–37

Zaman V (1997) Phase-contrast microscopy of cell division in *Blastocystis hominis*. Ann Trop Med Parasitol 91:223–224

Zaman V, Ng GC, Suresh K, Yap EH, Singh M (1993) Isolation of *Blastocystis* from the cockroach (Dictyoptera: Blattidae). Parasitol Res 79:73–74

Zaman V, Khan KZ (1994) A concentration technique for obtaining viable cysts of *Blastocystis hominis* from faeces. J Pak Med Assoc 44:220–221

Zaman V, Howe J, Ng M (1995) Ultrastructure of *Blastocystis hominis* cysts. Parasitol Res 81:465–469

Zaman V, Howe J, Ng M, Goh TK (1999) Scanning electron microscopy of the surface coat of *Blastocystis hominis*. Parasitol Res 85:974–976

Zhang X, Qiao JY, Zhou XJ, Yao FR, Wei ZC (2007) Morphology and reproductive mode of *Blastocystis hominis* in diarrhea and in vitro. Parasitol Res 101:43–51

Zhang X, Zhang S, Qiao J, Wu X, Zhao L, Liu Y, Fan X (2011) Ultrastructural insights into morphology and reproductive mode of *Blastocystis hominis*. Parasitol Res 110(3):1165–1172

Zierdt CH (1973) Studies of *Blastocystis hominis*. J Protozool 20:114–121

Zierdt CH (1988) *Blastocystis hominis*, a long-misunderstood intestinal parasite. Parasitol Today 4:15–17

Zierdt CH (1991) *Blastocystis hominis*—past and future. Clin Microbiol Rev 4:61–79

Chapter 3
Epidemiology, Transmission, and Zoonotic Potential of *Blastocystis* in Human and Animals

Hisao Yoshikawa

Abstract The genus *Blastocystis* is one of the most frequently encountered protozoans in human and animal fecal samples. In general, prevalence of *Blastocystis* infection in humans is higher in developing countries than in developed countries. However, detectability of the parasite is closely related to the diagnostic methods. Namely, routine parasitological diagnostic methods including direct fecal smear with or without staining and formalin–ether concentration methods are known to be considerably lower detection or almost undetectable. Therefore, suitable methods should be used for diagnosis of this parasite. The actual mode of transmission of *Blastocystis* between human-to-human, human-to-animal, and animal-to-animal is not conclusively demonstrated. However, the direct or indirect transmission of the parasite via fecal–oral route has been demonstrated between individuals in the closed human communities and among experimental animals in the same gages. In addition, only cyst form is known to be responsible for transmission. Therefore, contamination of fecal materials in drinking water or foods seems to be the most possible route for infection. Indeed, viable *Blastocystis* cysts have been found in the source of drinking water. Although zoonotic potential of the parasite has been proposed based on the genetic similarities among *Blastocystis* isolates from humans and animals, positive infections of some human isolates in rats and chickens indicate the zoonotic feature of this parasite. However, there are variability in the infectivity of the isolates in relation to genetic diversity of *Blastocystis*.

H. Yoshikawa (✉)
Department of Biological Sciences, Nara Women's University, Nara, Japan
e-mail: h.yoshikawa@cc.nara-wu.ac.jp

H. Mehlhorn et al. (eds.), *Blastocystis: Pathogen or Passenger?*, Parasitology Research Monographs 4,
DOI 10.1007/978-3-642-32738-4_3,

3.1 Introduction

Distribution of *Blastocystis* is worldwide in humans and in a variety of animals. However, the presence of *Blastocystis* had been reported in leeches and flies at 1910 years; these reports cannot be verified because of the problems in descriptions and resolution in the light microscope (Zierdt 1991). In this section, therefore, the prevalence of *Blastocystis* will focus on the infallible epidemiological reports.

3.2 Epidemiology

Considerable numbers of the reports are now available for prevalence of *Blastocystis* infection in human and animal. Prevalence of human *Blastocystis* infection is different among countries or among different regions in the same country. In general, however, prevalence of *Blastocystis* is higher in developing countries than in developed countries (Table 3.1). The higher prevalence in developing countries is due to the poor sanitary infrastructure and/or low level of care to environmental hygiene, because this parasite is known to be infected by the ingestion of the cyst form via fecal–oral route (Yoshikawa et al. 2004e; Tanizaki et al. 2005; Leelayoova et al. 2008). Although detectability of the parasite is correlated with diagnostic method, the traditional parasitological examinations, such as formalin–ether concentration method, wet-mount fresh fecal smear or Lugol's iodine-staining, and trichrome staining are considerably less sensitive (Suresh and Smith 2004; Termmathurapoj et al. 2004; Yakoob et al. 2004; Dogruman-Al et al. 2010). Indeed, the reported data examined with fecal smear, formalin–ether concentration method, or trichrome staining are considerably lower than those examined by in vitro cultures among the same countries (Table 3.1). The short-term in vitro culture assay is now gold standard for the diagnosis of *Blastocystis* infection in humans and animals because of the highest detection rate among various stool examinations (Stensvold et al. 2009c; Dogruman-Al et al. 2010). Although the culture assay requires at least 2 days for growth, it is also cost effective, and increasing the number of organisms does not require expertise of laboratory diagnosticians.

High prevalence in human infection of the parasite was observed in developing countries, even using routine diagnostic technique in epidemiological survey (Table 3.1). These results indicate that real prevalence of these countries is higher than that of reported data. Although the prevalence in developed countries is generally lower than developing countries, several countries such as Turkey (32.5 %) and USA (22.9 %) show relatively high prevalence. In contrast, the prevalence in Singapore (3.3 %) and Japan (2.5 %) is very low among developed countries even using the most sensitive method of in vitro culture.

There are several reports on different prevalence of the parasite among different regions or populations in the same countries. When the prevalence of *Blastocystis*

Table 3.1 Prevalence of *Blastocystis* infection in human populations in various countries

Country	Number of examined/positive (%)	Method for diagnosis	References
Developing countries			
Iran	4,371/254 (5.8 %)	Smear/FEC	Arani et al. (2008)
Jordan	500/73 (14.6 %)	Lugol's iodine/trichrome	Nimri (1993)
Argentina	504/137 (27.2 %)	Telemann technique	Basualdo et al. (2007)
China	703/26 (3.7 %)	Smear	Wang et al. (2002)
	239/78 (32.6 %)	In vitro culture	Li et al. (2007b)
Egypt	168/56 (33.3 %)	Trichrome	Rayan et al. (2007)
Thailand	1,920/119 (6.2 %)	FEC	Ngrenngarmlert et al. (2007)
	814/110 (13.5 %)	In vitro culture	Yaicharoen et al. (2006)
	904/334 (36.9 %)	In vitro culture	Leelayoova et al. (2002)
Philippines	172/70 (40.7 %)	FEC	Baldo et al. (2004)
Malaysia	130/68 (52.3 %)	Trichrome	Noor Azian et al. (2007)
Zambia	93/50 (53.8 %)	Lugol's iodine	Graczyk et al. (2005)
Indonesia	348/208 (59.8 %)	Formalin/smear	Pegelow et al. (1997)
Chile	970/599 (61.8 %)	PAF/sedimentation method	Torres et al. (1992)
Developed countries			
Turkey	2,047/317 (15.5 %)	Smear/FEC/trichrome	Dagci et al. (2008)
	197/64 (32.5 %)	In vitro culture	Dogruman-Al et al. (2008)
United States	2,896/662 (22.9 %)	CONSED™	Amin (2002)
Germany	67/12 (17.9 %)	In vitro culture	Yoshikawa et al. (2004d)
Italy	5,351/378 (7.1 %)	Smear/trichrome/WIH	Masucci et al. (2011)
Spain	8,313/585 (7.0 %)	MIF/smear	González-Moreno et al. (2011)
England	1,390/96 (6.9 %)	SAF/trichrome	Windsor et al. (2002)
Singapore	276/9 (3.3 %)	In vitro culture	Wong et al. (2008)
Japan	2,037/50 (2.5 %)	In vitro culture	Yoshikawa et al. (2004d)

FEC formalin–ether concentration method; *PAF* phenol–alcohol–formalin fixation; *WIH* Weigert's iron hemotoxylin; *MIF* merthiolate–iodine–formalin; *SAF* sodium acetate–acetic acid–formalin fixation

was investigated by culture method at four distinct areas among two distant western and eastern China, the prevalence of the parasite in Yunnan province in western China was several fold higher than that in Shanghai municipality and Yongjia country in eastern China (Li et al. 2007a). Significant differences of the prevalence between two regions in China are described as the difference of socio-economic development between inland province of Yunnan and coastal provinces of Shanghai and Zhejiang because the economic development of China started in the coastal provinces. In the survey in Thailand by in vitro culture, prevalence of *Blastocystis* infections in two different populations of schoolchildren showed 13.5 % (Yaicharoen et al. 2006) and 18.9 % (Leelayoova et al. 2008), respectively, while the army population indicated the prevalence of 36.9 % (Leelayoova et al. 2002). The considerably higher prevalence in the soldier population was significantly correlated to the consuming unfiltered or unboiled water.

Table 3.2 Prevalence of *Blastocystis* infection in animals

Animals	Prevalence (country)	References
Mammals		
Pigs	46.8 % (Spain)	Navarro et al. (2008)
	60.0 % (Great Britain)	Burden et al. (1978/1979)
	82.7 % (Czechoslovakia)	Pakandl (1991)
	95.1 % (Japan)	Abe et al. (2002)
	96.1 % (Thailand)	Thathaisong et al. (2003)
Cattle	1.8 % (Spain)	Quílez et al. (1995)
	70.9 % (Japan)	Abe et al. (2002)
Zoo primates	2.1 % (Malaysia)	Lim et al. (2008)
	53.8 % (Australia)	Parkar et al. (2010)
	66.6 % (Spain)	Pérez Cordón et al. (2008)
	57.7–85.3 % (Japan)	Yamada et al. (1987), Abe et al. (2002)
Wild primates	16.0 % (Ethiopia)	Legesse and Erko (2004)
Pet macaques	43.2 % (Indonesia)	Jones-Engel et al. (2004)
Dogs	36.1 % (Chile)	López et al. (2006)
	21.7–70.8 % (Australia)	Duda et al. (1998), Parkar et al. (2007)
Cats	37.4 % (Chile)	López et al. (2006)
	67.3 % (Australia)	Duda et al. (1998)
Rats	60.0 % (Singapore)	Chen et al. (1997)
Birds		
Chickens	74.4 % (Australia)	Lee and Stenzel (1999)
	100 % (Japan)	Yamada et al. (1987)
Ostriches	58.0 % (Spain)	Ponce Gordo et al. (2002)
Zoo pheasants	80.0 % (Japan)	Abe et al. (2002)
Zoo ducks	56.3 % (Japan)	Abe et al. (2002)
Reptiles	28.6 % (Singapore)	Teow et al. (1992)
Amphibians	41.1 % (Japan)	Yoshikawa et al. (2004b)
Cockroaches	80 % (Singapore)	Zaman et al. (1993)

Blastocystis organism is also commonly isolated from a variety of members of the Animal Kingdom, including mammals, birds, reptiles, amphibians, and insects. The prevalence of *Blastocystis* in animals greatly varies among different animal species or among different countries in the same animal species (Table 3.2). In mammals, domestic pigs and cattle, and zoo, wild, or pet primates are commonly infected with *Blastocystis*. In birds, domestic chickens and ostriches, and zoo pheasants and ducks are also frequently infected with *Blastocystis*. In zoo and pet animals, however, prevalence of *Blastocystis* is variable among the countries. For example, *Blastocystis* was isolated from giraffes and elephants in Australia, Belgium, and The Netherlands, while these animals were negative in Japan (Abe et al. 2002; Parkar et al. 2010). *Blastocystis* infection in dogs and cats were common in Chile and Australia, while it was not detected in Japan (Duda et al. 1998; Abe et al. 2002; López et al. 2006; Parkar et al. 2007). In laboratory animals, rats were highly infected with *Blastocystis*, while mice were totally negative (Chen et al. 1997). *Blastocystis* was also isolated from reptiles such as snakes, iguanas, and tortoises from zoo animals in Singapore (Teow et al. 1992). In amphibians, *Blastocystis* was common in the wild frogs, while it was totally negative in the wild newts in Japan (Yoshikawa et al. 2004b). In insects, *Blastocystis* is commonly isolated from sewer cockroaches in Singapore (Zaman et al. 1993). Therefore, a unique unicellular parasite, *Blastocystis*, may be one of the morphologically indistinguishable organisms adapted widely among the Animal Kingdom.

3.3 The Mode of Transmission

The actual mode of transmission of *Blastocystis* infection among wild animals and between human-to-human or animal-to-human is not conclusively demonstrated. Molecular approach on human *Blastocystis* infections in a small closed-community demonstrated direct human-to-human transmission of a certain genotype (subtype, ST) of *Blastocystis* among patients and between patients and staff members (Yoshikawa et al. 2000). Namely, *Blastocystis* ST3 had spread widely among patients and staff members in a facility A, the patients of another facility B were also infected by the same ST3 after two patients of the facility A moved to facility B. Several molecular studies also showed implication with contaminated water and close contact with pets and firm or wild animals as being a source of *Blastocystis* infections (Taamasri et al. 2000; Leelayoova et al. 2008; Stensvold et al. 2009b; Yoshikawa et al. 2009; Eroglu and Koltas 2010). These reports support the direct or indirect transmission of *Blastocystis* between individuals via fecal-oral route similar to some intestinal protozoan, including *Cryptosporidium* spp., *Entamoeba* spp., *Giardia intestinalis*, and *Dientamoeba fragilis*. These parasites are actually frequently coinfected with *Blastocystis* (Amin 2002; Yaicharoen et al. 2006; Ngrenngarmlert et al. 2007; Noor Azian et al. 2007; Leelayoova et al. 2008; Mehraj et al. 2008; Nuchprayoon et al. 2009; Souppart et al. 2009; Stensvold et al. 2009b; Masucci et al. 2011). The direct transmission of *Blastocystis* is also suggested by screening of the environmental factors (Basualdo et al. 2007). When many environmental factors were investigated in the rural population in Argentina, commensal amoebas, *G. lamblia* and *Entamoeba coli* were detected in the samples of soil and consumption water, while *Blastocystis* was only one case in the water samples, although the prevalence of *Blastocystis* was several times higher than other protozoan. Therefore direct transmission is reasonable for *Blastocystis* infection than indirect transmission in this survey.

Mode of transmission and the form of the parasite to contribute infection are recently well characterized by using animal models of rats and chickens. A series of oral inoculation of cultured organisms or cyst form revealed that only cyst form was an infective stage (Yoshikawa et al. 2004e). Fecal–oral transmission was also demonstrated in both rats and chickens from an infected one to an uninfected one when both animals had been housed in the same cage (Yoshikawa et al. 2004e; Tanizaki et al. 2005). Interestingly, infection of the uninfected control rat and chicken was recognized at 9 days and 4 days, respectively.

Infectivity of the concentrated cyst from the infected animal stool was also well characterized. Administration of ten cysts into three groups of ten rats showed the different infectivity from 10 to 100 % (Yoshikawa et al. 2004e), suggesting that contaminated water and food with a few number of *Blastocystis* cysts can establish an infection. The finding of viable cysts in Scottish and Malaysian sewage samples and in surface water in Malaysia also supports the possible transmission by drinking water (Suresh et al. 2005; Ithoi et al. 2011). Molecular epidemiological studies also indicate that consumption of unboiled or raw water plants are infective sources of *Blastocystis* (Taamasri et al. 2000; Leelayoova et al. 2004, 2008; Li et al. 2007b). Several studies showed that cyst form could survive in water for up to 19 days or 1 month at room temperature or 25 °C, respectively, and for 2 months at 4 °C (Moe et al. 1997; Yoshikawa et al. 2004e). These differences may be strain or isolate difference, but more studies are required to determine life span of the cyst form in natural environment. Therefore, only the cyst form sustains transmission of *Blastocystis* infection via direct or indirect fecal–oral route.

3.4 Zoonotic Potential

The first study suggesting zoonotic potential of *Blastocystis* revealed susceptibility of human isolates in germ-free guinea pigs via oral and intracecal inoculation of the cultured organisms (Phillip and Zierdt 1976). This study showed the presence of the organisms up to 120 days in a 3-week-old animal. The later studies used germ-free piglets or SPF mice for experimental infection of human isolates (Pakandl et al. 1993; Moe et al. 1997). Interestingly, *Blastocystis* organisms were only found in one piglet sacrificed at 4 days post-inoculation, while in mice infections were observed up to 2 weeks post-inoculation. These results are most likely to due to low susceptibility of those animals, although *Blastocystis* infection in guinea pigs and piglets are commonly observed. Subsequently, zoonotic potential of *Blastocystis* isolates from various vertebrate hosts have been considered based on the sequencing of the SSU rRNA gene (Arisue et al. 2003; Noël et al. 2003, 2005; Thathaisong et al. 2003; Yoshikawa et al. 2004c; Scicluna et al. 2006; Parkar et al. 2007; Navarro et al. 2008; Rivera 2008; Stensvold et al. 2009a) or various PCR-based methodologies (Yoshikawa et al. 1996, 2003, 2004a; Clark 1997; Abe et al. 2003a,b,c; Abe 2004; Rivera and Tan 2005; Yan et al. 2007). If *Blastocystis* can transmit among a variety of mammals and birds including humans, we are facing is how to assign species name for these *Blastocystis* isolates from humans and animals (Yoshikawa et al. 2004c). In 2007, therefore, *Blastocystis* isolates from mammalian and avian animal hosts has been consensually named *Blastocystis* sp. and it classifies all isolates into 9 STs based on the phylogeny inferred with SSU rRNA gene (Stensvold et al. 2007b). Because some reptilian and amphibian isolates are positioned within the ranges of variation by the mammalian and avian clades (Yoshikawa et al. 2004c; Noël et al. 2005), if these isolates are distinct species, all of the nine clades (nine STs) cannot be *B. hominis*. So far, all nine STs of

Blastocystis have been identified among human populations, while some STs are limited among the isolates from mammals or birds, or specific animal group. The recent accumulated molecular data indicate that the majority of human isolates were ST3 in most countries (Yoshikawa et al. 2004d; Scicluna et al. 2006; Stensvold et al. 2006, 2007a, 2009a; Li et al. 2007a,b; Dogruman-Al et al. 2008, 2009; Menounos et al. 2008; Özyurt et al. 2008; Wong et al. 2008; Rene et al. 2009; Souppart et al. 2009, 2010; Meloni et al. 2011). Although a few showed that the major genotype of human isolates was ST1 or ST4 rather than ST3 (Thathaisong et al. 2003; Leelayoova et al. 2008; Domínguez-Márquez et al. 2009), in the studies nearly half of the isolates were unidentified ST (Thathaisong et al. 2003). Conversely, only a few isolates from animals without primates are found to belong to ST3 (Abe et al. 2003b; Yoshikawa et al. 2004a; Stensvold et al. 2009a). Interestingly, all these animal ST3 isolates are found to be from firm or pet animals, which are potentially in close contact with humans. These evidences suggest that only a genetically distinct genotype (ST3) of *Blastocystis* can easily infect humans, whereas other genotypes are sporadic or more suitable to animal hosts. Indeed, experimental infection in rats and chickens with human *Blastocystis* isolates of ST3 was fully unsuccessful (Iguchi et al. 2007), while some other subtypes are consistent to the evidence of natural survey. Namely, in avian ST6 and ST7, three out of four human STs 6 and 7 isolates could infect only chickens, while the remaining one isolate could infect both chickens and rats. Conversely, both of two human isolates of rodent ST4 could infect chickens, while only one isolate could infect rats. It is reasonable to recognize that some STs are fully host specific, while others are in progress adaptation to the hosts.

Several circumstance evidences also support the zoonotic feature of *Blastocystis*. High prevalence rate among food handlers, animal handlers, and abattoir workers are observed (Sadek et al. 1997; Salim et al. 1999; Requena et al. 2003; Khan and Alkhalife 2005; Parkar et al. 2007, 2010), indicating that certain occupations such as close contact to animals is high risk of infection with zoonotic *Blastocystis*. The finding of human *Blastocystis* infections with ST5 in the rural area in China also shows possible infection with the zoonotic ST5 parasites from pigs because ST5 is predominant in farm pigs and cattle (Yan et al. 2007). Moreover, molecular epidemiological studies in the local survey areas or specific occupation also support the transmission of the specific STs from animals or pets to humans (Parkar et al. 2007; Yoshikawa et al. 2009; Stensvold et al. 2009b; Eroglu and Koltas 2010). Therefore, more detail researches are needed to validate relationship between host specificity and *Blastocystis* STs,and zoonotic potential of genetically polymorphic parasite.

3.5 Conclusion

The genus *Blastocystis* is probably one of the most indistinguishable eukaryotic unicellular organisms found in a variety animals including humans, whether these isolates are genetically identically or represent unique species is presently unclear.

Unfortunately, any current molecular methodologies developed are insufficient for identification or classification of genetically different organisms, because the genus *Blastocystis* is extensively genetically polymorphic. Therefore new molecular tools for these purposes should be developed for this parasite. Recently, there have been increased evidences of zoonotic potential of this parasite and which genotype(s) (subtype(s)) is correlated with pathogenicity is still under debate. Although there are few evidences of the fecal-oral transmission via cyst form of *Blastocystis*, there are only limited data are available for existence of the cyst form in the drinking water or natural environment. Therefore, it should be investigated to what extent such potentially zoonotic subtypes are correlated with pathogenicity, and clearly, more studies should aim at clarifying issues related to transmission routes and host specificity.

References

Abe N, Nagoshi M, Takami K, Sawano Y, Yoshikawa H (2002) A survey of *Blastocystis* sp. in livestock, pets, and zoo animals in Japan. Vet Parasitol 106:203–212

Abe N, Wu Z, Yoshikawa H (2003a) Molecular characterization of *Blastocystis* isolates from birds by PCR with diagnostic primers and restriction fragment length polymorphism analysis of the small subunit ribosomal RNA gene. Parasitol Res 89:393–396

Abe N, Wu Z, Yoshikawa H (2003b) Zoonotic genotypes of *Blastocystis hominis* detected in cattle and pigs by PCR with diagnostic primers and restriction fragment length polymorphism analysis of the small subunit ribosomal RNA gene. Parasitol Res 90:124–128

Abe N, Wu Z, Yoshikawa H (2003c) Molecular characterization of *Blastocystis* isolates from primates. Vet Parasitol 113:321–325

Abe N (2004) Molecular and phylogenetic analysis of *Blastocystis* isolates from various hosts. Vet Parasitol 120:235–242

Amin OM (2002) Seasonal prevalence of intestinal parasites in the United States during 2000. Am J Trop Med Hyg 66:799–803

Arani AS, Alaghehbandan R, Akhlaghi L, Shahi M, Lari AR (2008) Prevalence of intestinal parasites in a population in south of Tehran, Iran. Rev Inst Med Trop S Paulo 50:145–149

Arisue N, Hashimoto T, Yoshikawa H (2003) Sequence heterogeneity of the small subunit ribosomal RNA genes among *Blastocystis* isolates. Parasitology 126:1–9

Baldo ET, Belizario VY, De Leon WU, Kong H-H, Chung D-I (2004) Infection status of intestinal parasites in children living in residential institutions in Metro Manila, the Philippines. Korean J Parasitol 42:67–70

Basualdo JA, Córdoba MA, De Luca MM, Ciarmela ML, Pezzani BC, Grenovero MS, Minvielle MC (2007) Intestinal parasitoses and environmental factors in a rural population of Argentina, 2002–2003. Rev Inst Med Trop S Paulo 49:251–255

Burden DJ, Anger HS, Hammet NC (1978/1979) *Blastocystis* sp. infection in pigs. Vet Microbiol 3:227–234

Chen XQ, Singh M, Ho LC, Moe KT, Tan SW, Yap EH (1997) A survey of *Blastocystis* sp. in rodents. Lab Anim Sci 47:91–94

Clark CG (1997) Extensive genetic diversity in *Blastocystis hominis*. Mol Biochem Parasitol 87:79–83

Dagci H, Kurt Ö, Demirel M, Östan I, Azizi NR, Mandiracioglu A, Yurdagül C, Tanyüksel M, Eroglu E, Ak M (2008) The prevalence of intestinal parasites in the province of Izmir, Turkey. Parasitol Res 103:839–845

Dogruman-Al F, Dagci H, Yoshikawa H, Kurt Ö, Demirel M (2008) A possible link between subtype 2 and asymptomatic infections of *Blastocystis hominis*. Parasitol Res 103:685–689
Dogruman-Al F, Yoshikawa H. Kustimur S, Balaban N (2009) PCR-based subtyping of *Blastocystis* isolates from symptomatic and asymptomatic individuals in a major hospital in Ankara, Turkey. Parasitol Res 106:263–268
Dogruman-Al F, Simsek Z, Boorom K, Ekici E, Sahin M, Tuncer C, Kustimur S, Altinbas A (2010) Comparison of methods for detection of *Blastocystis* infection in routinely submitted stool samples, and also in IBS/IBD patients in Ankara, Turkey. PLoS One 5:e15484
Domínguez-Márquez MV, Guna R, Muñoz C, Gómez-Muñoz T, Borrás R (2009) High prevalence of subtype 4 among isolates of *Blastocystis hominis* from symptomatic patients of a health district of Valencia (Spain). Parasitol Res 105:949–955
Duda A, Stenzel DJ, Boreham PF (1998) Detection of *Blastocystis* sp. in domestic dogs and cats. Vet Parasitol 76:9–17
Eroglu F, Koltas IS (2010) Evaluation of the transmission mode of *B. hominis* by using PCR method. Parasitol Res 107:841–845
González-Moreno O, Domingo L, Teixidor J, Gracenea M (2011) Prevalence and associated factors of intestinal parasitisation: a cross-sectional study among outpatients with gastrointestinal symptoms in Catalonia, Spain. Parasitol Res 108:87–93
Graczyk TK, Shiff CK, Tamang L, Munsaka F, Beitin AM, Moss WJ (2005) The association of *Blastocystis hominis* and *Endolimax nana* with diarrheal stools in Zambian school-age children. Parasitol Res 98:38–43
Iguchi A, Ebisu A, Nagata S, Saitou Y, Yoshikawa H, Iwatani S, Kimata I (2007) Infectivity of different genotypes of human *Blastocystis hominis* isolates in chickens and rats. Parasitol Int 56:107–112
Ithoi I, Jali A, Mak JW, Sulaiman WYW, Mahmud R (2011) Occurrence of *Blastocystis* in water of two rivers from recreational areas in Malaysia. J Parastiol Res 2011: Article ID 123916
Jones-Engel L, Engel GA, Schillaci MA, Kyes K, Froehlich J, Paputungan U, Kyes RC (2004) Prevalence of enteric parasites in pet macaques in Sulawesi, Indonesia. Am J Primatol 62:71–82
Khan ZA, Alkhalife IS (2005) Prevalence of *Blastocystis hominis* among "healthy" food handlers in Dammam, Soudi Arabia. J Egypt Soc Parasitol 35:395–401
Lee MG, Stenzel DJ (1999) A survey of *Blastocystis* in domestic chickens. Parasitol Res 85:109–117
Leelayoova S, Taamasri P, Rangsin R, Naaglor T, Thathaisong U, Mungthin M (2002) In-vitro cultivation: a sensitive method for detecting *Blastocystis hominis*. Annal Trop Med Parasitol 96:803–807
Leelayoova S, Rangsin R, Taamasri P, Naaglor T, Thathaisong U, Mungthin M (2004) Evidence of waterborne transmission of *Blastocystis hominis*. Am J Trop Med Hyg 70:658–662
Leelayoova S, Siripattanapipong S, Thathaisong U, Naaglor T, Taamasri P, Piyaraj P, Mungthin M (2008) Drinking water: a possible source of *Blastocystis* spp. subtype 1 infection in schoolchildren of a rural community in central Thailand. Am J Trop Med Hyg 79:401–406
Legesse M, Erko B (2004) Zoonotic intestinal parasites in *Papio anubis* (baboon) and *Cercopithecus aethiops* (vervet) from four localities in Ethiopia. Acta Trop 90:231–236
Li LH, Zhang X-P, Lv S, Zhang L, Yoshikawa H, Wu Z, Steinmann P, Utzinger J, Tong X-M, Chen S-H, Zhou X-N (2007a) Cross-sectional surveys and subtype classification of human *Blastocystis* isolates from four epidemiological setting in China. Parasitol Res 102:83–90
Li LH, Zhou X-N, Du Z-W, Wang X-Z, Wang L-B, Jiang J-Y, Yoshikawa H, Steinman P, Utzinger J, Wu Z, Chen J-X, Chen S-H, Zhang L (2007b) Molecular epidemiology of human *Blastocystis* in a village in Yunnan province, China. Parasitol Int 56:281–286
Lim YAL, Ngui R, Shukri J, Rohela M, Mat Naim HR (2008) Intestinal parasites in various animals at a zoo in Malaysia. Vet Parasitol 157:154–159
López JD, Abarca KV, Paredes PM, Inzunza ET (2006) Intestinal parasites in dogs and cats with gastrointestinal symptoms in Santiago, Chile. Rev Med Chile 134:193–200 (In Portuguese)

Masucci L, Graffeo R, Bani S, Bugli F, Boccia S, Nicolotti N, Fiori B, Fadda G, Spanu T (2011) Intestinal parasites isolated in a large teaching hospital, Italy, 1 May 2006 to 31 December 2008. Euro Surveill 16(24), pii: 19891

Mehraj V, Hatcher J, Akhtar S, Rafique G, Beg MA (2008) Prevalence and factors associated with intestinal parasitic infection among children in an urban slum of Karachi. PLoS One 3:e3680

Meloni D, Sanciu G, Poirier P, El Alaoui H, Chabé M, Delhaes L, Dei-Cas E, Delbac F, Fiori PL, Di Cave D, Viscogliosi E (2011) Molecular subtyping of *Blastocystis* sp. isolates from symptomatic patients in Italy. Parasitol Res 109:613–619

Menounos PG, Spanakos G, Tegos N, Vassalos CM, Papadopoulou C, Vakalis NC (2008) Direct detection of *Blastocystis* sp. in human faecal samples and subtype assignment using single strand conformational polymorphism and sequencing. Mol Cell Probes 22:24–29

Moe KT, Singh M, Howe J, Ho LC, Tan SW, Chen XQ, Ng GC, Yap EH (1997) Experimental *Blastocystis hominis* infection in laboratory mice. Parasitol Res 83:319–325

Navarro C, Domínguez-Márquez MV, Garijo-Toledo MM, Vega-García S, Fernández-Barredo S, Pérez-Gracia MT, García A, Borrás R, Gómez-Muñoz MT (2008) High prevalence of *Blastocystis* sp. in pigs reared under intensive growing system: frequency of ribotypes and associated risk factors. Vet Parasitol 153:347–358

Ngrenngarmlert W, Lamom C, Pasuralertsakul S, Yaicharoen R, Wongjindanon N, Sripochang S, Suwajeejarun T, Sermsart B-O, Kiatfuengfoo R (2007) Intestinal parasitic infections among school children in Thailand. Trop Biomed 24:83–88

Nimri LF (1993) Evidence of an epidemic of *Blastocystis h*ominis infections in preschool children in northern Jordan. J Clin Microbiol 31:2706–2708

Noël C, Peyronnet C, Gerbod D, Edgcomb VP, Delgado-Viscogliosi P, Sogin ML, Capron M, Viscogliosi E, Zenner L (2003) Phylogenetic analysis of *Blastocystis* isolates from different hosts based on the comparison of small-subunit rRNA gene sequences. Mol Biochem Parasitol 126:119–123

Noël C, Dufernez F, Gerbod D, Edgcomb VP, Delgado-Viscogliosi P, Ho L-C, Singh M, Wintjens R, Sogin ML, Capron M, Pierce R, Zenner L, Viscogliosi E (2005) Molecular phylogenies of *Blastocystis* isolates from different hosts: implications for genetic diversity, identification of species, and zoonosis. J Clin Microbiol 43:348–355

Noor Azian MY, San YM, Gan CC, Yusri MY, Nurulsyamzawaty Y, Zuhaizam AH, Maslawaty MN, Norparina I, Vythilingam I (2007) Prevalence of intestinal protozoa in an aborigine community in Pahang, Malaysia. Trop Biomed 24:55–62

Nuchprayoon S, Sanprasert V, Kaewzaithim S, Saksirisampant W (2009) Screening for intestinal parasitic infections among Myanmar migrant workers in Thai food industry: a high-risk transmission. J Immigrant Minority Health 11:115–121

Özyurt M, Kurt Ö, Mølbak K, Nielsen HV, Haznedaroglu T, Stensvold CR (2008) Molecular epidemiology of *Blastocystis* infections in Turkey. Parasitol Int 57:300–306

Pakandl M (1991) Occurrence of *Blastocystis* sp. in pigs. Folia Parasitol 38:297–301

Pakandl M, Koudela B, Vítovec J (1993) An experimental infection of conventional and gnotobiotic piglets with human and porcine strains of *Blastocystis*. Folia Parasitol 40:319–320

Parkar U, Traub RJ, Kumar S, Mungthin M, Vitali S, Leelayoova S, Morris K, Thompson RCA (2007) Direct characterization of *Blastocystis* from faeces by PCR and evidence of zoonotic potential. Parasitolology 134:359–367

Parkar U, Traub RJ, Vitali S, Elliot A, Levecke B, Robertson I, Geurden T, Steele J, Drake B, Thompson RCA (2010) Molecular characterization of *Blastocystis* isolates from zoo animals and their animal-keepers. Vet Parasitol 169:8–17

Pegelow K, Gross R, Pietrzik K, Lukito W, Richards AL, Fryauff DJ (1997) Parasitological and nutritional situation of school children in the Sukaraja district, West Java, Indonesia. Southeast Asia J Trop Med Public Health 28:173–190

Pérez Cordón G, Hitos Prados A, Romero D, Sánchez Moreno M, Pontes A, Osuna A, Rosales MJ (2008) Intestinal parasitism in the animals of the zoological garden "Peña Escrita" (Almuñecar, Spain). Vet Parasitol 156:302–309

Phillip BP, Zierdt CH (1976) *Blastocystis hominis*: pathogenic potential in human patients and in gnotobiotes. Exp Parasitol 39:358–364

Ponce Gordo F, Herrera S, Castro AT, García Durán B, Martínez Díaz RA (2002) Parasites from farmed ostriches (*Struthio camelus*) and rheas (*Rhea americana*) in Europe. Vet Parasitol 107:137–160

Quílez J, Sánchez-Acedo C, Clavel A, Causapé AC (1995) Occurrence of *Blastocystis* sp. in cattle in Aragón, northeastern Spain. Parasitol Res 81:703–705

Rayan HZ, Ismail OA, El Gayar EK (2007) Prevalence and clinical features of *Dientamoeba fragilis* infections in patients suspected to have intestinal parasitic infection. J Egypt Soc Parasitol 37:599–608

Requena I, Hernández Y, Ramsay M, Salazar C, Devera R (2003) Prevalence of *Blastocystis hominis* among food handlers from Caroni municipality, Bolivar State, Venezuela. Cad Saude Publica Rio de Janeiro 19:1721–1727

Rene BA, Stensvold CR, Badsberg JH, Nielsen HV (2009) Subtype analysis of *Blastocystis* isolates from *Blastocystis* cyst excreting patients. Am J Trop Med Hyg 80:588–592

Rivera WL (2008) Phylogenetic analysis of *Blastocystis* isolates from animal and human hosts in the Philippines. Vet Parasitol 156:178–182

Rivera WL, Tan MA (2005) Molecular characterization of *Blastocystis* isolates in the Philippines by riboprinting. Parasitol Res 96:253–257

Sadek Y, el-Fakahany AF, Lashin AH, el-Salam FA (1997) Intestinal parasites among food-handlers in Qualyobia Governorate, with reference to the pathogenic parasite *Blastocystis hominis*. J Egypt Soc Parasitol 27:471–478

Salim HR, Kumar GS, Vellayan S, Mak JW, Anuar AK, Init I, Vennila GD, Saminathan R, Ramakrishnan K (1999) *Blastocystis* in animal handlers. Parasitol Res 85:1032–1033

Scicluna SM, Tawari B, Clark CG (2006) DNA barcoding of *Blastocystis*. Protist 157:77–85

Souppart L, Sanciu G, Cian A, Wawrzyniak I, Delbac F, Capron M, Dei-Cas E, Boorom K, Delhaes L, Viscogliosi E (2009) Molecular epidemiology of human *Blastocystis* isolates in France. Parasitol Res 105:413–421

Souppart L, Moussa H, Cian A, Sanciu G, Poirier P, Al Alaoui H, Delbac F, Boorom K, Delhaes L, Dei-Cas E, Viscogliosi E (2010) Subtype analysis of *Blastocystis* isolates from symptomatic patients in Egypt. Parasitol Res 106:505–511

Stensvold R, Brillowska-Dabrowska A, Nielsen HV, Arendrup MC (2006) Detection of *Blastocystis hominis* in unpreserved stool specimens by using polymerase chain reaction. J Parasitol 92:1081–1087

Stensvold CR, Arendrup MC, Jespersgaard C, Mølbak K, Nielsen HV (2007a) Detecting *Blastocystis* using parasitologic and DNA-based methods: a comparative study. Diagn Micribiol Infect Dis 59:303–307

Stensvold CR, Suresh GK, Tan KSW, Thompson RCA, Traub RJ, Viscogliosi E, Yoshikawa H, Clark CG (2007b) Terminology for *Blastocystis* subtypes—a consensus. Trends Parasitol 23:93–96

Stensvold CR, Alfellani MA, Nørskov-Lauritsen S, Prip K, Victory EL, Maddox C, Nielsen HV, Clark CG (2009a) Subtype distribution of *Blastocystis* isolates from synanthropic and zoo animals and identification of a new subtype. Int J Parasitol 39:473–479

Stensvold CR, Lewis HC, Hammerum AM, Porsbo LJ, Mielsen SS, Olsen KEP, Arendrup MC, Nielsen HV, Mølbak K (2009b) *Blastocystis*: unravelling potential risk factors and clinical significance of a common but neglected parasite. Epidemiol Infect 137:1655–1663

Stensvold CR, Nielsen HV, Mølbak K, Smith HV (2009c) Pursuing the clinical significance of *Blastocystis*—diagnostic limitations. Trends Parasitol 25:23–29

Suresh K, Smith H (2004) Comparison of methods for detecting *Blastocystis hominis*. Eur J Clin Microbiol Infect Dis 23:509–511

Suresh K, Smith HV, Tan TC (2005) Viable *Blastocystis* cysts in Scottish and Malaysian sewage samples. Appl Environ Microbiol 71:5619–5620

Taamasri P, Mungthin M, Rangsin R, Tongupprakarn B, Areekul W, Leelayoova S (2000) Transmission of intestinal blastocystosis related to the quality of drinking water. Southeast Asian J Trop Med Public Health 31:112–117

Tanizaki A, Yoshikawa H, Iwatani S, Kimata I (2005) Infectivity of *Blastocystis* isolates from chickens, quails and geese in chickens. Parasitol Res 96:57–61

Teow WL, Ng GC, Chan PP, Chan YC, Yap EH, Zaman V, Singh M (1992) A survey of *Blastocystis* in reptiles. Parasitol Res 78:453–455

Termmathurapoj S, Leelayoova S, Aimpun P, Thathaisong U, Nimmanon T, Taamasri P, Mungthin M (2004) The usefulness of short-term in vitro cultivation for the detection and molecular study of *Blastocystis hominis* in stool specimens. Parasitol Res 93:445–447

Thathaisong U, Worapong J, Mungthin M, Tan-Ariya P, Viputtigul K, Sudatis A, Noonai A, Leelayoova S (2003) *Blastocystis* isolates from a pig and a horse are closely related to *Blastocystis hominis*. J Clin Microbiol 41:967–975

Torres P, Miranda JC, Flores L, Riquelme J, Franjola R, Perez J, Auad S, Hermosilla C, Riquelme S (1992) Blastocystosis and other intestinal protozoan infections in human riverside communities of the Valdivia River Basin, Chile. Rev Inst Med Trop S Paulo 34:557–564 (In Portuguese)

Wang K-X, Li C-P, Wang J, Cui Y-B (2002) Epidemiological survey of *Blastocystis hominis* in Huainan city, Anhui province, China. World J Gastrointerol 8:928–932

Windsor JJ, Macfarlane L, Hughes-Thapa G, Jones SK, Whiteside TM (2002) Incidence of *Blastocystis hominis* in faecal samples submitted for routine microbiological analysis. Br J Biomed Sci 59:154–157

Wong KHS, Ng GC, Lin RTP, Yoshikawa H, Taylor MB, Tan KSW (2008) Predominance of subtype 3 among *Blastocystis* isolates from a major hospital in Singapore. Parasitol Res 102:663–670

Yaicharoen R, Ngrenngarmlert W, Wongjindanon N, Sripochang S, Kiatfuengfoo R (2006) Infection of *Blastocystis hominis* in primary schoolchildren from Nakhon Pathom province, Thailand. Trop Biomed 23:117–122

Yakoob J, Jafri W, Jafri N, Khan R, Islam M, Asim Beg M, Zaman V (2004) Irritable bowel syndrome: in search of an etiology: role of *Blastocystis hominis*. Am J Trop Med Hyg 70:383–385

Yamada M, Yoshikawa H, Tegoshi T, Matsumoto Y, Yoshikawa T, Shiota T, Yoshida Y (1987) Light microscopical study of *Blastocystis* spp. in monkeys and fowls. Parasitol Res 73:527–531

Yan Y, Su S, Ye J, Lai X, Lai R, Liao H, Chen G, Zhang R, Hou Z, Luo X (2007) *Blastocystis* sp. subtype 5: a possibly zoonotic genotype. Parasitol Res 101:1527–1532

Yoshikawa H, Nagano I, Yap EH, Singh M, Takahashi Y (1996) DNA polymorphism revealed by arbitrary primers polymerase chain reaction among *Blastocystis* strains isolated from humans, a chicken, and a reptile. J Eukaryot Microbiol 43:127–130

Yoshikawa H, Abe N, Iwasawa M, Kitano S, Nagano I, Wu Z, Takahashi Y (2000) Genomic analysis of *Blastocystis hominis* strains isolated from two long-term health care facilities. J Clin Microbiol 38:1324–1330

Yoshikawa H, Wu Z, Nagano I, Takahashi Y (2003) Molecular comparative studies among *Blastocystis* isolates obtained from humans and animals. J Parasitol 89:585–594

Yoshikawa H, Abe N, Wu Z (2004a) PCR-based identification of zoonotic isolates of *Blastocystis* from mammals and birds. Microbiology 150:1147–1151

Yoshikawa H, Morimoto K, Nagashima M, Miyamoto N (2004b) A survey of *Blastocystis* infection in anuran and urodele amphibians. Vet Parasitol 122:91–102

Yoshikawa H, Morimoto K, Wu Z, Singh M, Hashimoto T (2004c) Problems in speciation in the genus *Blastocystis*. Trends Parasitol 20:251–255

Yoshikawa H, Wu Z, Kimata I, Iseki M, Ali IKMD, Hossain MB, Zaman V, Haque R, Takahashi Y (2004d) Polymerase chain reaction-based genotype classification among human *Blastocystis hominis* populations isolated from different countries. Parasitol Res 92:22–29

Yoshikawa H, Yoshida K, Nakajima A, Yamanari K, Iwatani S, Kimata I (2004e) Fecal-oral transmission of the cyst form of *Blastocystis hominis* in rats. Parasitol Res 94:391–396

Yoshikawa H, Wu Z, Pandy K, Pandy BD, Sherchand JB, Yanagi T, Kanbara H (2009) Molecular characterization of *Blastocystis* isolates from children and rhesus monkeys in Kathmandu, Nepal. Vet Parasitol 160:295–300

Zaman V, Ng GC, Suresh K, Yap EH, Singh M (1993) Isolation of *Blastocystis* from the cockroach (Dictyoptera: Blattidae). Parasitol Res 79:73–74

Zierdt CH (1991) *Blastocystis hominis*—past and future. Clin Microbiol Rev 4:61–79

Chapter 4
Blastocystis–Host Interactions: Insights from In Vitro Model Systems

Kevin S.W. Tan and Haris Mirza

Abstract The enteric protistan parasite *Blastocystis* is an extracellular luminal organism of controversial clinical relevance. In the absence of an established animal model to study pathogenesis, in vitro systems provide some clues to the pathobiology of *Blastocystis*. Such studies using specific *Blastocystis* subtypes have shown that the parasite is able to induce a variety of effects on intestinal epithelial cell lines including barrier compromise, cell death, and the production of proinflammatory cytokines. Other studies also provide some evidence that the parasite is able to evade the host immune response by suppressing iNOS production and cleaving immunoglobulins. Membrane insert (transwell) systems for the study of *Blastocystis*–host interactions are useful platforms for investigating inter- and intra-subtype variations in pathogenesis.

Keywords *Blastocystis* • Host–parasite interactions • Barrier function • Immune response • In vitro • Intestinal epithelial cells • IEC • Subtypes • Pathogenesis

4.1 Introduction

The compilation of a book on *Blastocystis* is timely as 2011 marks the 100th anniversary since the first descriptions of this enigmatic parasite was reported (Alexeieff 1911; Brumpt 1912). As time progressed, it became increasingly clear

K.S.W. Tan (✉)
Laboratory of Molecular and Cellular Parasitology, Department of Microbiology, Yong Loo Lin School of Medicine, National University of Singapore, 5 Science Drive 2, Singapore 117597, Singapore
e-mail: mictank@nus.edu.sg

H. Mirza
Yale University School of Medicine, 333 Cedar Street, New Haven, CT 06510, USA
e-mail: haris.mirza@yale.edu

H. Mehlhorn et al. (eds.), *Blastocystis: Pathogen or Passenger?*, Parasitology Research Monographs 4,
DOI 10.1007/978-3-642-32738-4_4, © Springer-Verlag Berlin Heidelberg 2012

that *Blastocystis* was a divisive organism. Disagreements on its life cycle, modes of reproduction, phylogenetic affiliations, and clinical relevance plagued the Parasitology field for decades. While it is currently accepted that *Blastocystis* is transmitted via the fecal–oral route, reproduces by binary fission and is classified as a stramenopile, its clinical relevance is still highly controversial (Tan 2008). There are a number of reasons for this. Firstly, *Blastocystis* can be commonly found in both healthy and symptomatic individuals. Secondly, symptoms attributed to blastocystosis are generally mild and non specific. Thirdly, it has been recently appreciated that *Blastocystis* exists as a species complex, and the genus comprises numerous genetically distinct subtypes (ST) (Stensvold et al. 2007), each of which exert varying levels of cytopathic effects on host cells. Lastly, there is an absence of a good animal model to study the pathobiology of the parasite. This deficiency will likely be overcome in the coming years as interest in the parasite increases. In the absence of a relevant animal model to study *Blastocystis* infections, in vitro culture systems provide useful alternatives to study *Blastocystis*–host interactions. In this chapter, we will discuss how such and related studies have shed light on *Blastocystis* pathogenic mechanisms and also provide some evidence for how the parasite may evade the host immune response.

4.2 In Vitro Assays That Mimic *Blastocystis*–Intestinal Epithelial Cell Interactions

Membrane insert systems, also referred to as transwell systems, are in-well tissue culture devices that mimic the in vivo environment by allowing the laboratory culture of cell monolayers with apical and basolateral exposure to nutrients. These permeable supports have polycarbonate membrane filters with pore sizes ranging from 0.1 to 12 μm. Designed for the culture of polarized cells, cellular activities can be studied readily because cells grown on filters provide convenient, independent access to apical and basal membrane domains (Fig. 4.1). Supports are usually available as 6-, 12- and 24-well plates and in a dish format. Applications include studies on transport and permeability, cell invasion, migration, chemotaxis, and cellular control and differentiation. A number of studies using classical animal models support the in vitro findings obtained using membrane insert systems, highlighting their physiological relevance. Due to their high reproducibility and low cost they can also be readily adapted for large-scale screening studies. Inserts in various formats are commercially available from a number of sources (Millipore Millicell®, Corning Transwell®, BD Biosciences BD Falcon™ Insert Systems).

In the context of *Blastocystis*–host studies, membrane insert platforms have been applied to barrier function, tight junction, and immunological assays (Mirza et al. 2011b; Mirza et al. 2012; Puthia et al. 2006; Teo et al. unpublished observations). The format of the assay allows for both contact-dependent and contact-independent studies to be carried out. In the former, intestinal epithelial cell (IEC) monolayers

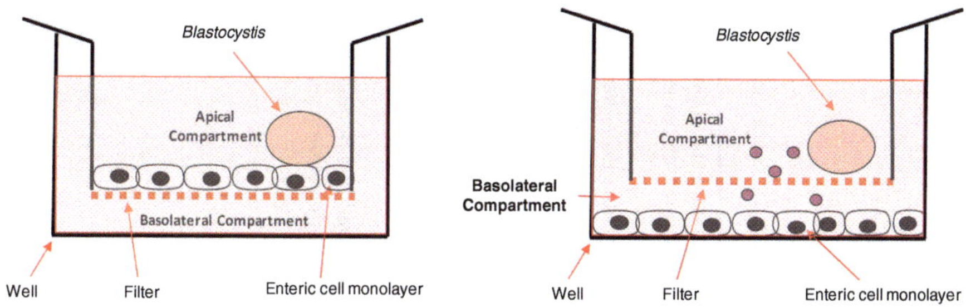

Fig. 4.1 Diagrammatic representations of a typical membrane insert system for *Blastocystis*–host studies. *Left*, this setup allows the growth of IEC monolayers on permeable membrane supports and is recommended for coculture, in vitro toxicology, permeability and electrophysiology experiments. IEC monolayers can be exposed to *Blastocystis* cells, extracts or purified components within the apical compartment. IEC cells can be assayed for apoptosis, cytokine production and barrier function disruption. *Right*, IEC can also be grown on the well base, and *Blastocystis* can be added to the apical compartment to study contact-independent effects

are cultured on the surface of the membrane support before the addition of *Blastocystis* cells onto the apical chamber. For the latter assay, IECs are grown on the bottom well of the basolateral compartment while parasite cells are introduced into the apical chamber, allowing only molecules smaller than the size of the membrane pores to interact with the IECs (Fig. 4.1).

Various IECs have been used to study *Blastocystis*–host interactions (Table 4.1). The nontransformed rat epithelial cell line IEC-6 was used to study cytopathic effects of *Blastocystis* ST4 (Puthia et al. 2006). In separate studies, the human colon adenocarcinoma cell line HT-29 was used for cytotoxic assays after exposure to *Blastocystis* ST1 (Long et al. 2001) or clinical isolates (Walderich et al. 1998) but there was a lack of significant cytopathic effect on this cell line. The human colonic carcinoma cell line T-84 also showed a lack of cytopathic effect when exposed to *Blastocystis* ST1, but produced proinflammatory cytokine IL-8 upon exposure to the parasite (Long et al. 2001). A later study using the same cell line, but upon exposure to a ST4 isolate also revealed that T-84 cell lines produced IL-8 when exposed to parasite lysate or live cells (Puthia et al. 2008). Caco-2, a heterogeneous human epithelial colorectal adenocarcinoma cell line, is perhaps the most versatile IEC for the study of host–pathogen interactions as numerous relevant assays can be performed on these cells. On membrane inserts and on coverslips, the cells differentiate into polarized monolayers and express tight junctions, microvilli and are capable of secreting cytokines (Hou et al. 2010; Humen et al. 2011). Caco-2 cells have been employed for apoptosis (unpublished observations), cytokine (unpublished observations), tight junction (Mirza et al. 2012), and nitric oxide (NO) studies (Mirza et al. 2011b) on *Blastocystis*. *Blastocystis* readily adheres to Caco-2 monolayers (Wu et al. unpublished), and such interactions can be visualized by confocal differential interference contrast microscopy (Fig. 4.2) and confocal fluorescence microscopy (Fig. 4.3). Table 4.2 summarizes studies on various *Blastocystis* isolates and subtypes on IECs.

Table 4.1 Mammalian cell lines employed for *Blastocystis*–host studies

Cell line	Origin	Properties	Applications
Caco-2	Human, epithelial colorectal adenocarcinoma	Polarized on monolayers, express tight junctions, microvilli, enzymes and transporters characteristic of enterocytes	Barrier function studies, drug absorption and efflux studies, NO studies, cytokine studies
HT-29	Human, epithelial colorectal adenocarcinoma	Undifferentiated under standard growth conditions, sparse microvilli, low brush border enzyme activities under standard growth conditions	Studies of factors involved in differentiation of epithelial cells, cytokine and apoptosis studies
T-84	Human, epithelial colorectal carcinoma	Polarized on monolayers, express tight junctions and desmosomes, express receptors for many peptide hormones and neurotransmitters, and maintain vectorial electrolyte transport	Barrier function studies, studies on epithelial chloride secretion, apoptosis, mucin and cytokine production
HCT-116	Human, epithelial colorectal carcinoma	Nonpolarized and undifferentiated, positive for keratin and transforming growth factor beta 1 (TGFβ-1) and beta 2 (TGFβ-2) expression	Studies of factors involved in differentiation of epithelial cells, growth and drug sensitivity studies
IEC-6	Rat (*Rattus norvegicus*), small intestine, epithelial, non-transformed crypt-like cell line	Polarized on monolayers, express tight junctions, microvilli	Barrier function, absorption, cytokine and apoptosis studies

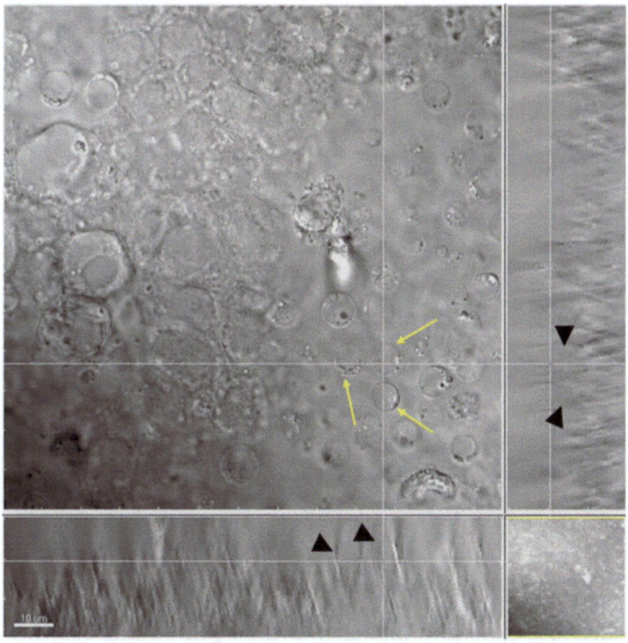

Fig. 4.2 Confocal differential interference contrast micrographs of *Blastocystis* sp. ST7 in contact with Caco-2 cell lines. Typical *Blastocystis* cells (*yellow arrows*) are seen among larger Caco-2 epithelial cells. The micrograph corresponds to an optical plane at the *top* of the epithelial monolayer. In XZ and YZ transverse sections, the parasites are shown in close proximity (*black arrowheads*) to the apical regions of the epithelial cells

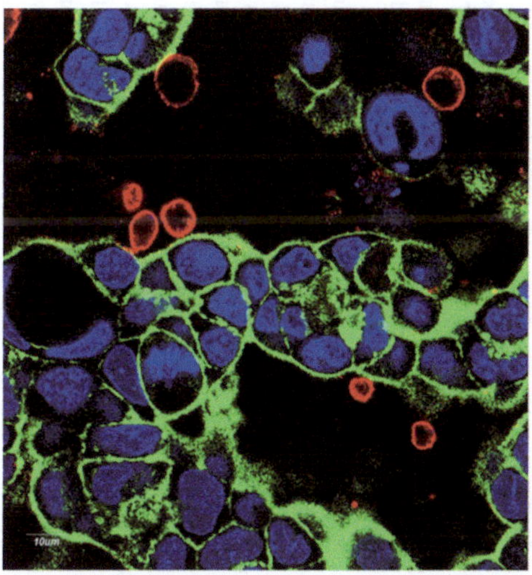

Fig. 4.3 Fluorescence confocal micrographs of *Blastocystis* sp. ST7 in contact with Caco-2 cell lines. The micrograph corresponds to an optical plane at the *top* of the epithelial monolayer *Blastocystis* sp. ST7 is labeled with *Blastocystis* legumain antibody-mAb1D5, a murine IgM monoclonal antibody and visualized with secondary Alexa Fluor® 594 goat anti-mouse IgM (*red*). Caco-2 cells are stained with actin-specific Phalloidin-FITC (*green*) and DAPI (*blue*).

Table 4.2 *Blastocystis* species and subtypes employed for in vitro host-pathogen studies

Species	Isolate	Subtype	Remarks	References
B. hominis	Nand II	1	Lack of cytopathic effects but induction of proinflammatory responses on HT-29 and T-84 cell lines	Long et al. (2001)
B. hominis	B	7	Isolated from a symptomatic patient. Proteases cleave sIgA, and induce IL-8 production from T-84 cell line. Proteases also mediate tight junction disruption in Caco-2 cells. Parasites suppress iNOS gene transcription in Caco-2 cells. High arginase activity and competes with IECs for L-arginine to suppress NO production. Metronidazole resistant	Mirza et al. (2011b), Mirza et al. (2012), Puthia et al. (2005), Puthia et al. (2008)
B. ratti	WR-1	4	Proteases cleave sIgA, and compromise barrier function and induces apoptosis in IEC-6 cells	Puthia et al. (2005), Puthia et al. (2006)
B. hominis	Clinical isolates from healthy and symptomatic carriers	Unknown	Cells and lysates induced significant cytopathic effects on CHO but not in HT-29 cells	Walderich et al. (1998)
B. hominis	Clinical isolates from healthy and symptomatic carriers	Unknown	Lysates induced proliferation of HCT-116 cells with lysates from symptomatic patients induced greater increase in cell proliferation. These lysates also induced upregulation of NF-κB, IL-6 and IL-8	Chan et al. (2012)

4.3 Effects of *Blastocystis* on Intestinal Epithelial Cell Barrier Function

Intestinal barrier function is mainly regulated by tight junctions, which are the most apical component of the epithelial junctional complex and are vital for the selective diffusion of ions and solutes along the paracellular pathway and for restricting luminal pathogens and/or their products access to underlying tissue (Marchiando et al. 2010). Tight junctions consist of transmembrane proteins such as claudins, occluding and junctional adhesion molecules that provide cell–cell contact, and cytoplasmic peripheral membrane proteins, including zonula-occludens-1, -2, and -3, and cingulin, that anchor the tight junction complex to the actin cytoskeleton (Clayburgh et al. 2004).

Numerous gastrointestinal disorders are associated with a breakdown of epithelial barrier function. These include bacterial enteritis, celiac disease, and inflammatory bowel disease (Marchiando et al. 2010). Evidence for *Blastocystis*-mediated barrier compromise was reported by Puthia et al. (2006). Using membrane insert systems, it was reported that *Blastocystis ratti* (ST4) mediated barrier dysfunction evidenced by decrease in transepithelial resistance and increase in transepithelial flux in IEC-6 cells. Pretreatment with caspase inhibitors failed to rescue *Blastocystis*-induced changes, and the authors suggested that apoptosis was not a major factor contributing to the loss of barrier function. This contrasts with *Giardia*, whereby parasite-induced enterocyte apoptosis was associated with barrier compromise (Chin et al. 2002). In the case of *Blastocystis*, more subtypes and IECs need to be tested before a clearer picture on the role of *Blastocystis*-mediated apoptosis and barrier dysfunction emerges. We have recently observed that cysteine proteases of *Blastocystis* disrupt Caco-2 monolayer tight junctions in a Rho-kinase (ROCK)-mediated fashion resulting in colonic epithelial barrier compromise. The pathological outcomes are mitigated when components of the Rho/ROCK pathways are blocked by the inhibitors Fasudil and Simvastatin (Mirza et al. 2012). This is the first study highlighting the cytoprotective effect of a statin in colonic epithelial pathology mediated by parasite proteases. Additionally, the study suggests that statins may be administered to blastocystosis patients as combination or adjunctive therapy in order to protect the host epithelia from parasite-mediated pathology.

4.4 Effects of *Blastocystis* on Host Defenses

At mucosal surfaces of the gastrointestinal tract, secretory immunoglobulin A (sIgA) is the major immune defense against microbial pathogens and their toxins. sIgA functions to limit mucosal adhesion and colonization of pathogens and also serve to neutralize toxins. Ig A proteases, enzymes that cleave Ig A molecules, have been reported for luminal parasites such as *Blastocystis* (Puthia et al. 2006), *Trichomonas vaginalis* (Provenzano and Alderete 1995) and *Entamoeba histolytica*

(Quezada-Calvillo and López-Revilla 1987). In the *Blastocystis* study, both spent media and cell lysates from subtypes 4 and 7 parasites were able to degrade human sIgA in vitro as evidenced by fragmentation and decrease in intensity of IgA bands in western blots (Puthia et al. 2006). Inhibition studies revealed that cysteine proteases from a ST7 isolate and aspartic proteases from a ST4 isolate mediate IgA degradation. The study suggests that *Blastocystis* proteases are virulence factors and contribute to survival by degrading neutralizing mucosal antibodies. It would be interesting to investigate if *Blastocystis* from symptomatic patients possess greater immunogloblulin-degrading ability when compared to isolates from asymptomatic patients, as has been shown for *T. vaginalis* (Provenzano and Alderete 1995).

Nitric oxide is a multifunctional molecule that plays important roles in the human body, such as maintenance of vascular tone, modulation of epithelial barrier function, and neurotransmission. NO also has antimicrobial properties and is produced by many cell types in response to pathogens. The subsequent nitrosative stress is an important aspect of innate immune defense against pathogenic microbes. Protozoan parasites have evolved a variety of mechanisms to evade the host NO response. *Entamoeba histolytica* consumes L-arginine, a major cellular substrate for NO production (Elnekave et al. 2003), thus limiting levels of this amino acid for iNOS-mediated NO production. In contrast, *Toxoplasma* and *Leishmania* decrease NO levels by downregulation of iNOS mRNA (Rozenfeld et al. 2005; Matte and Descoteaux 2010). A recent study investigated the role of NO in *Blastocystis*–epithelial interactions (Mirza et al. 2011b). The authors compared the susceptibility of metronidazole-resistant (MZ^r) and -sensitive (MZ^s) strains of *Blastocystis* to nitrosative stress. Interestingly, the MZ^r strain was more sensitive to NO donors when compared to a MZ^s strain. Using a Caco-2 in vitro membrane insert system, the authors observed, by real-time PCR, that the MZ^r but not the MZ^s strain was able to inhibit NO production by downregulating the inducible nitric oxide synthase (iNOS) gene. They surmised that MZ^r strains of *Blastocystis* suffer a fitness cost at the expense of their drug resistance molecular machinery. This is compensated by the ability of the parasite to decrease host NO levels by inhibiting the mRNA levels of the iNOS gene. This is the first study to reveal strain to strain variation in the ability of *Blastocystis* in modulating the host innate immune response and may shed light on the possible mechanisms employed by *Blastocystis* to persist in the host.

Collectively, the current studies suggest that *Blastocystis* is able to modulate the host immune response, by mechanisms that include degradation of secretory immunoglobulins and suppression of host NO production. Such suppression may also subvert these protective responses to other coinfecting pathogens, allowing these microbes to gain a foothold on the host epithelia and cause disease. This "bystander effect" by *Blastocystis* would be an interesting area of investigation in the context of coinfections.

4.5 Effects of *Blastocystis* on Host Immune Response

There have only been a handful of studies focusing on the host immune response to *Blastocystis*. Despite this, experimental animal infection and in vitro studies support the notion that *Blastocystis* is able to induce a proinflammatory immune response. Earlier studies using ST7-infected Balb/c mice revealed intense infiltration of inflammatory cells into the colon and inflammation with edematous lamina propria in the cecum and colon (Moe et al. 1997). In contrast, in a later study using a rat model (Iguchi et al. 2009), infection with a ST4 isolate revealed neither mucosal sloughing nor inflammatory cell infiltration but showed an increase in goblet cell numbers. Real-time PCR quantification revealed an upregulation of the expression of type interferon-γ, interleukin (IL)-12, and tumor necrosis factor alpha, but not IL-6 or granulocyte-macrophage colony-stimulating factor. The authors concluded that the parasite induced a proinflammatory response that was insufficient to result in local tissue damage. The discrepancies in the pathological outcomes may be attributed to subtype- and host-dependent effects. Balb/c mice are not natural hosts for ST7 and experimental infections led to an acute infectious course (Moe et al. 1997) while the ST4 infection of rats, their natural hosts, resulted in a chronic infection (Iguchi et al. 2009).

Using an in vitro experimental system, *Blastocystis* Nand II (ST 1) parasites were reported to elicit a significant increase in proinflammatory chemokines IL-8 and granulocyte-macrophage colony-stimulating factor (GM-CSF) in HT-29 and T-84 IEC human colonic cell lines (Long et al. 2001). This was observable after 24 h exposure to parasites. However, at 6 h, the parasite exposure resulted in a suppression of *Escherichia coli*- and LPS-induced production of IL-8. The authors suggested that *Blastocystis* is able to modulate the host immune response depending on the phase of infection. At earlier stages of infection, the parasite downregulates the host immune response for survival. In a later study (Puthia et al. 2008), T-84 cell lines exposed to *Blastocystis ratti* (ST4) live cells and lysates revealed significant expression of IL-8 chemokine in an NF-κB-dependent manner. IL-8 production was diminished upon cysteine protease inhibiton, suggesting a role for *Blastocystis* proteases in virulence. The authors suggested that in vivo, *Blastocystis* infection mediates IEC production of IL-8, causing an influx of inflammatory cells into the intestinal mucosa with resultant tissue damage and gastrointestinal disturbances.

Future studies must include comparisons across different *Blastocystis* subtypes and IECs, because it is difficult to make any conclusions based on studies utilizing single cell types. Since the outcome of an immune response depends on the balance between pro- and anti-inflammatory responses, it would be necessary to also conduct assays that provide a comprehensive overview of cytokine profiles including the anti-inflammatory cytokines IL-4, -10 and -13.

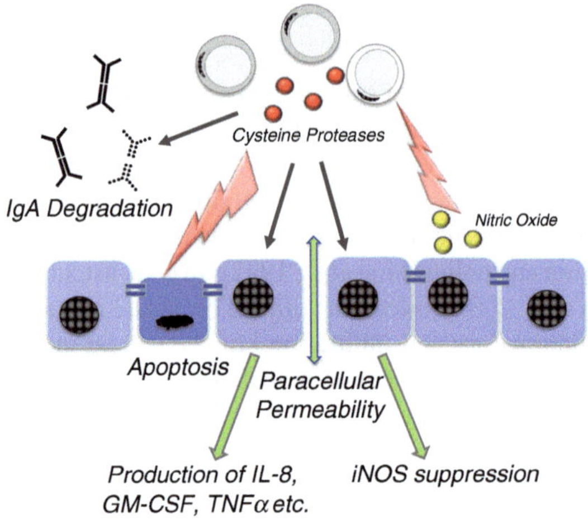

Fig. 4.4 A model for *Blastocystis* pathogenesis at the cellular level. *Blastocystis* infection may result in a variety of pathological outcomes such as secretory IgA degradation, barrier function compromise via alterations to tight junctions, host cell apoptosis, and induction of proinflammatory cytokines IL-8, GM-CSF and TNFα. IgA degradation and barrier disruption may promote the growth and invasion of neighboring pathogens. Parasite cysteine proteases have been shown to mediate the majority of these features. *Blastocystis* has also been shown to inhibit host NO response by transcriptional inhibition of the iNOS gene

4.6 Progress, Challenges, and Future Directions

Blastocystis research has accelerated in the last decade. Notably, the genome of a ST7 isolate (B) was recently published (Denoeud et al. 2011), providing researchers with an invaluable resource for molecular and genetic studies. Analysis of the genome revealed that *Blastocystis* produces all major classes of proteases, with cysteine proteases being the most represented. Among the cysteine protease genes, 96 % are predicted to be secreted. Interestingly, *Blastocystis* secretes five types of the cysteine protease legumain, an unusual asparaginyl endopeptidase, which has been known to play roles in degradation of extracellular matrix proteins and in activating pro-enzymes. *Blastocystis* legumain was also shown to be localized to the parasite surface and was important for the survival of the parasite (Wu et al. 2010). With the genome information available, reverse genetic studies can be carried out to understand the roles of specific putative virulence factors. A challenge to achieving this is the lack of a workable transfection system for *Blastocystis* to facilitate gene knock-out and expression studies. The author is confident that this will change in the not-too-distant future.

Figure 4.4 summarizes our current understanding of *Blastocystis*–host interactions from in vitro studies. The studies described in this chapter are too few and much more work needs to be done for one to make firm conclusions on the pathogenesis of *Blastocystis*. This is especially pertinent because *Blastocystis* exhibits inter- and intra-subtype variations in its biology (Mirza and Tan 2009;

Mirza et al. 2011a,b). Thus, more isolates need to be used on a variety of both primary and transformed cell lines so that a comprehensive picture of *Blastocystis* virulence can be achieved. This will certainly be a great challenge considering the amount of work required and the dearth of research funding for neglected diseases such as blastocystosis. A relevant animal model is urgently needed for in vitro studies to be validated. Rats and chicks have been used with some success (Yoshikawa et al. 2004; Tanizaki et al. 2005). It has been proposed that mice are unlikely to be good animal models since mice do not naturally harbor *Blastocystis* and do not exhibit long lasting experimental infections (Tan 2008). However, most studies are conducted on mice from a single genetic background (Balb/c) and it is well accepted that inbred mice from different genetic backgrounds can display significant variations in susceptibility to the same pathogen, as has been shown for *Entamoeba* (Hamano et al. 2008). Future studies should therefore investigate the infectivity of *Blastocystis* in inbred mice of different genetic backgrounds, such as C57BL/6, C3H, CBA mice. Mice that display severe or chronic manifestations of blastocystosis will be invaluable tools to study its little-understood pathogenesis.

There are still major gaps in our understanding of the basic biology of *Blastocystis*. For example, nothing yet is known about the adhesion of *Blastocystis* to host cells, although this is the first crucial step in the pathogenic process. The in vitro systems described in this chapter would be useful tools for the development of an adhesion assay for *Blastocystis*, as has been described for *E. histolytica* (Seigneur et al. 2005). Subsequently, the molecular basis for adhesion and inhibition studies can be carried out.

There have been interesting reports that *Blastocystis* is able to mediate pathology beyond intestinal dysfunction. Infection with *Blastocystis* has been linked to cutaneous disorders, such as urticaria (Zuel-Fakkar et al. 2011) while parasite lysate was observed to promote the growth of colorectal tumor cell line HCT 116 (Chandramathi et al. 2010; Chan et al. 2012). Future studies should focus on the mechanistic basis for the unusual outcomes of these *Blastocystis*–host interactions.

With the plethora of cutting-edge cell and molecular biology tools available today, one can hope that our understanding of *Blastocystis*–host interactions would increase dramatically in the coming years. The missing ingredients for a successful recipe are adequate research funding and, more importantly, a critical mass of researchers with a passion for unraveling the mysteries surrounding *Blastocystis*.

References

Alexeieff A (1911) Sur la nature des formations dites kystes de *Trichomonas intestinalis*. C R Soc Biol 71:296–298

Brumpt E (1912) *Blastocystis hominis* n sp. et formes voisines. Bull Soc Pathol Exot 5:725–730

Chan KH, Chandramathi S, Suresh K, Chua KH, Kuppusamy UR (2012) Effects of symptomatic and asymptomatic isolates of *Blastocystis hominis* on colorectal cancer cell line, HCT116. Parasitol Res 110(6):2475–2480

Chandramathi S, Suresh K, Kuppusamy UR (2010) Solubilized antigen of *Blastocystis hominis* facilitates the growth of human colorectal cancer cells, HCT116. Parasitol Res 106(4):941–945
Chin AC, Teoh DA, Scott KG, Meddings JB, Macnaughton WK, Buret AG (2002) Strain-dependent induction of enterocyte apoptosis by *Giardia lamblia* disrupts epithelial barrier function in a caspase-3-dependent manner. Infect Immun 70(7):3673–3680
Clayburgh DR, Shen L, Turner JR (2004) A porous defense: the leaky epithelial barrier in intestinal disease. Lab Invest 84(3):282–291
Denoeud F, Roussel M, Noel B, Wawrzyniak I, Da Silva C, Diogon M, Viscogliosi E, Brochier-Armanet C, Couloux A, Poulain J, Segurens B, Anthouard V, Texier C, Blot N, Poirier P, Ng GC, Tan KSW, Artiguenave F, Jaillon O, Aury JM, Delbac F, Wincker P, Vivarès CP, El Alaoui H (2011) Genome sequence of the stramenopile *Blastocystis*, a human anaerobic parasite. Genome Biol 12(3):R29
Elnekave K, Siman-Tov R, Ankri S (2003) Consumption of L-arginine mediated by *Entamoeba histolytica* L-arginase (EhArg) inhibits amoebicidal activity and nitric oxide production by activated macrophages. Parasite Immunol 25(11–12):597–608
Hamano S, Becker S, Asgharpour A, Ocasio YP, Stroup SE, McDuffie M, Houpt E (2008) Gender and genetic control of resistance to intestinal amebiasis in inbred mice. Genes Immun 9(5):452–461
Hou Y, Mortimer L, Chadee K (2010) *Entamoeba histolytica* cysteine proteinase 5 binds integrin on colonic cells and stimulates NFkappaB-mediated pro-inflammatory responses. J Biol Chem 285(46):35497–35504
Humen MA, Pérez PF, Liévin-Le Moal V (2011) Lipid raft-dependent adhesion of *Giardia intestinalis* trophozoites to a cultured human enterocyte-like Caco-2/TC7 cell monolayer leads to cytoskeleton-dependent functional injuries. Cell Microbiol 13(11):1683–1702
Iguchi A, Yoshikawa H, Yamada M, Kimata I, Arizono N (2009) Expression of interferon gamma and proinflammatory cytokines in the cecal mucosa of rats experimentally infected with *Blastocystis* sp. strain RN94-9. Parasitol Res 105(1):135–140
Long HY, Handschack A, König W, Ambrosch A (2001) *Blastocystis hominis* modulates immune responses and cytokine release in colonic epithelial cells. Parasitol Res 87:1029–1030
Marchiando AM, Graham WV, Turner JR (2010) Epithelial barriers in homeostasis and disease. Annu Rev Pathol 5:119–144
Matte C, Descoteaux A (2010) *Leishmania donovani* amastigotes impair gamma interferon-induced STAT1alpha nuclear translocation by blocking the interaction between STAT1alpha and importin-alpha5. Infect Immun 78(9):3736–3743
Mirza H, Tan KSW (2009) *Blastocystis* exhibits inter- and intra-subtype variation in cysteine protease activity. Parasitol Res 104(2):355–361
Mirza H, Teo JDW, Upcroft J, Tan KSW (2011a) A rapid, high-throughput viability assay for *Blastocystis* spp. reveals metronidazole resistance and extensive subtype-dependent variations in drug susceptibilities. Antimicrob Agents Chemother 55:637–648
Mirza H, Wu Z, Kidwai F, Tan KSW (2011b) A metronidazole-resistant isolate of *Blastocystis* spp. is susceptible to nitric oxide and downregulates intestinal epithelial inducible nitric oxide synthase by a novel parasite survival mechanism. Infect Immun 79:5019–5026
Mirza H, Wu Z, Teo JDW, Tan KSW (2012) Statin pleiotropy prevents rho kinase-mediated intestinal epithelial barrier compromise induced by *Blastocystis* cysteine proteases. Cell Microbiol 14(9):1474–1484. doi:10.1111/j.1462-5822.2012.01814.x
Moe KT, Singh M, Howe J, Ho LC, Tan SW, Chen XQ, Ng GC, Yap EH (1997) Experimental *Blastocystis hominis* infection in laboratory mice. Parasitol Res 83(4):319–325
Provenzano D, Alderete JF (1995) Analysis of human immunoglobulin-degrading cysteine proteinases of *Trichomonas vaginalis*. Infect Immun 63(9):3388–3395
Puthia MK, Sio SW, Lu J, Tan KSW (2006) *Blastocystis ratti* induces contact-independent apoptosis, F-actin rearrangement, and barrier function disruption in IEC-6 cells. Infect Immun 74:4114–4123

Puthia MK, Vaithilingam A, Lu J, Tan KSW (2005) Degradation of human secretory immunoglobulin A by *Blastocystis*. Parasitol Res 97:386–389

Puthia MK, Lu J, Tan KSW (2008) *Blastocystis ratti* contains cysteine proteases that mediate interleukin-8 response from human intestinal epithelial cells in an NF-kappaB-dependent manner. Eukaryot Cell 7:435–443

Quezada-Calvillo R, López-Revilla R (1987) IgA protease in *Entamoeba histolytica* trophozoites. Adv Exp Med Biol 216B:1283–1288

Rozenfeld C, Martinez R, Seabra S, Sant'anna C, Gonçalves JG, Bozza M, Moura-Neto V, De Souza W (2005) *Toxoplasma gondii* prevents neuron degeneration by interferon-gamma-activated microglia in a mechanism involving inhibition of inducible nitric oxide synthase and transforming growth factor-beta1 production by infected microglia. Am J Pathol 167(4):1021–1031

Seigneur M, Mounier J, Prevost MC, Guillén N (2005) A lysine- and glutamic acid-rich protein, KERP1, from *Entamoeba histolytica* binds to human enterocytes. Cell Microbiol 7(4):569–579

Stensvold CR, Suresh GK, Tan KSW, Thompson RC, Traub RJ, Viscogliosi E, Yoshikawa H, Clark CG (2007) Terminology for *Blastocystis* subtypes—a consensus. Trends Parasitol 23:93–96

Tan KSW (2008) New insights on classification, identification and clinical relevance of *Blastocystis* spp. Clin Microbiol Rev 21(4):639–665

Tanizaki A, Yoshikawa H, Iwatani S, Kimata I (2005) Infectivity of *Blastocystis* isolates from chickens, quails and geese in chickens. Parasitol Res 96(1):57–61

Walderich B, Bernauer S, Renner M, Knobloch J, Burchard GD (1998) Cytopathic effects of *Blastocystis hominis* on Chinese hamster ovary (CHO) and adeno carcinoma HT29 cell cultures. Trop Med Int Health 3:385–390

Wu B, Yin J, Texier C, Roussel M, Tan KSW (2010) *Blastocystis* legumain is localized on the cell surface, and specific inhibition of its activity implicates a pro-survival role for the enzyme. J Biol Chem 285(3):1790–1798

Yoshikawa H, Yoshida K, Nakajima A, Yamanari K, Iwatani S, Kimata I (2004) Fecal-oral transmission of the cyst form of *Blastocystis hominis* in rats. Parasitol Res 94(6):391–396

Zuel-Fakkar NM, Abdel Hameed DM, Hassanin OM (2011) Study of *Blastocystis hominis* isolates in urticaria: a case-control study. Clin Exp Dermatol 36(8):908–910

Chapter 5
Clinical Aspects of *Blastocystis* Infections: Advancements Amidst Controversies

Haris Mirza and Kevin S.W. Tan

Abstract *Blastocystis* is a noninvasive, luminal parasite commonly reported in human stool samples. Its clinical presentation is diverse ranging between acute diarrhea and mild chronic abdominal discomfort. Clinical manifestations of *Blastocystis* also include urticaria and irritable bowel syndrome. Similar to other luminal parasites such as *Giardia* and *Entamoeba*, asymptomatic carriage of *Blastocystis* is common. The parasite is zoonotic and animal contact often leads to infections. The *Blastocystis* is also opportunistic with higher frequency in immunocompromised populations including pediatric and cancer patients as well as HIV-infected individuals. Although parasite is noninvasive it might complicate pathogenicity of other invasive pathogens. Metronidazole is the treatment of choice, but clinical management of *Blastocystis* is difficult owing to frequent reports of treatment failure. It also exhibits extensive phenotypic and genotypic diversity, not efficiently captured by classical diagnostic techniques, leading to under-reporting of *Blastocystis* infections. Diversity in pathobiology between variant parasite subtypes is suspected to be responsible for diverse clinical presentations of *Blastocystis* infections. Treatment outcomes are also suspected to be dependent on parasite subtype and phenotype. Despite a number of controversies surrounding the pathogenic potential of *Blastocystis*, several advances have been

H. Mirza (✉)
Yale University School of Medicine, 333 Cedar Street, New Haven, CT 06510, USA
e-mail: Haris.mirza@yale.edu

K.S.W. Tan
Laboratory of Molecular and Cellular Parasitology, Department of Microbiology, Yong Loo Lin School of Medicine, National University of Singapore, 5 Science Drive 2, Singapore 117597, Singapore
e-mail: mictank@nus.edu.sg

H. Mehlhorn et al. (eds.), *Blastocystis: Pathogen or Passenger?*,
Parasitology Research Monographs 4,
DOI 10.1007/978-3-642-32738-4_5, © Springer-Verlag Berlin Heidelberg 2012

made in recent years in the areas of parasite virulence as well as molecular and cellular biology. These advances have also fueled improvements in clinical diagnostic tools as well identification of novel treatment options. This chapter highlights some of these recent developments.

Keywords *Blastocystis* • Diagnosis • Diarrhea • Irritable bowel syndrome • Urticaria • Treatment • Resistance • Metronidazole • Subtypes • Pathogenesis

5.1 Introduction

Blastocystis is an extracellular, noninvasive, parasite colonizing the human colon (Tan et al. 2010). In several epidemiological studies, it is one of the most frequently isolated parasites with prevalence ranging between 5 % in industrialized countries and as high as 76 % in developed countries (Dogruman-Al et al. 2010; Tan 2008). The parasite exhibits extensive phenotypic and genotypic diversity. It has recently been classified into 13 subtypes (ST1-13), 9 of which are known to infect humans (Stensvold et al. 2007). Parasite subtypes display diverse pathobiology (Mirza and Tan 2009; Mirza et al. 2011a, b). Moreover, the parasite exists in four different morphological forms i.e., vacuolar, amoeboid, granular, and cystic, each exhibiting distinct characteristics (Tan et al. 2010). There is limited consensus on the clinical management of *Blastocystis* infections (Coyle et al. 2012). Limitations of diagnostic techniques, variable phenotypes and genotypes of *Blastocystis*, large number of asymptomatic carriers, and frequent reports of treatment failure are blamed for most of the controversies surrounding the clinical management of the parasite infection (Tan et al. 2010). This chapter describes the current advances in the clinical management of *Blastocystis* infections, addresses the controversies concerning the pathogenic potential of the parasite, and discusses the emerging clinical relevance of this mysterious microorganism.

5.2 Clinical Presentations

Symptomatic infections by the parasite are called blastocystosis (Tan et al. 2010) and these have diverse clinical presentations, ranging from self-limiting abdominal discomfort (Nagel et al. 2011; Stensvold et al. 2009c) to chronic persistent diarrhea recalcitrant to antiprotozoal therapy (Nagel et al. 2011; Stensvold et al. 2008). Frequently, the parasite infections also present with dermatological symptoms (Hameed et al. 2011; Zuel-Fakkar et al. 2011). The diverse symptomatology of the parasite is often explained by isolate or subtype-dependent variations in *Blastocystis* virulence (Hameed et al. 2011; Nagel et al. 2011).

5.2.1 Luminal Infections

The most common symptom of blastocystosis is diarrhea with abdominal pain, with varying severity (Tan et al. 2010). Other intestinal symptoms include nausea, vomiting, bloating anorexia, weight loss, constipation, and flatulence (Nagel et al. 2011; Stensvold et al. 2009c; Tan 2008; Tan et al. 2010). Although the parasite is frequently associated with intestinal symptoms, an unequivocal causal relationship between the *Blastocystis* and gastrointestinal disorders has not been established thus far primarily because of a large number of asymptomatic carriers. Consequently, Koch's postulates have not been demonstrated for *Blastocystis* (Coyle et al. 2012). The duration and severity of symptoms vary from acute enteritis to chronic mild diarrhea (Nagel et al. 2011; Stensvold et al. 2009c). There is no consensus on the possible reasons for variation in intestinal symptoms. Some studies associate these variations with infection density (Kaya et al. 2007; Moghaddam et al. 2005) while others implicate strain-dependent differences in virulence (Domínguez-Márquez et al. 2009; Eroglu et al. 2009; Mirza and Tan 2009; Mirza et al. 2011b; Nagel et al. 2011; Stensvold et al. 2009b, c). Interestingly some authors even suggest coinfection of *Blastocystis* with other known pathogens as a cause of intestinal symptoms (Herbinger et al. 2011).

5.2.1.1 Parasite Density

Some studies associate the severity of symptoms with *Blastocystis* density in stool samples (Kaya et al. 2007; Moghaddam et al. 2005). Higher frequency of acute intestinal symptoms were reported in patients with >5 parasites in each high-power field (×400) for wet mount or oil immersion (×1,000) in permanent stained smears (Kaya et al. 2007; Moghaddam et al. 2005). Recently in a case of life-threatening toxic megacolon, >10 *Blastocystis*/field were observed in the absence of *Clostridium difficile* (Salvi et al. 2012). Symptoms were resolved with intravenous metronidazole and trimethoprim–sulfamethoxazole treatment (Salvi et al. 2012).

5.2.1.2 Subtype-Dependent Pathogenicity

With the introduction of subtyping system for *Blastocystis* classification accumulating studies focus on association of specific parasite subtypes with human pathology (Nagel et al. 2011; Stensvold et al. 2009c). A number of reports suggest a strain to strain variation in parasite pathogenicity (Domínguez-Márquez et al. 2009; Eroglu et al. 2009; Mirza and Tan 2009; Mirza et al. 2011b; Nagel et al. 2011; Stensvold et al. 2009b, c). Studies associate subtype 1, 4, and 7 (Domínguez-Márquez et al. 2009; Eroglu et al. 2009; Nagel et al. 2011; Stensvold et al. 2009b, c) with human pathology whereas subtype 2 and 3 are considered nonpathogenic (Domínguez-Márquez et al. 2009; Eroglu et al. 2009; Stensvold et al. 2009c). However, mixed subtype infections

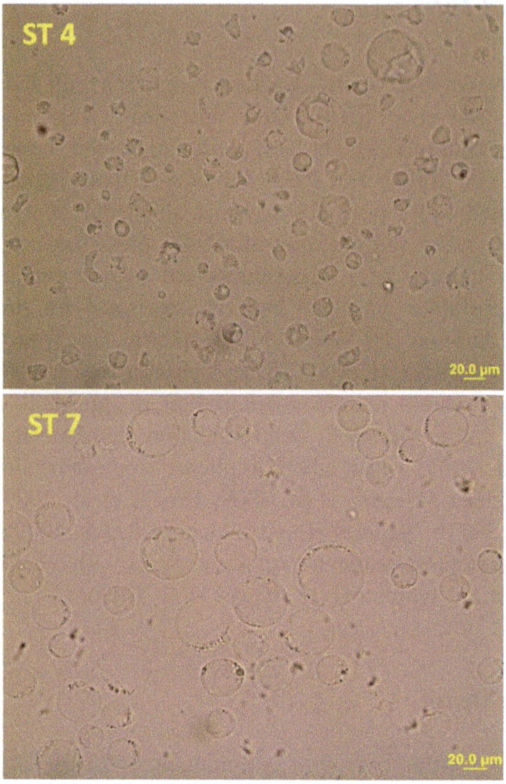

Fig. 5.1 Subtype-dependent morphological variations in *Blastocystis*. Micrographs representing morphological differences between vacuolar form of metronidazole-sensitive subtype 4 (*top*) and metronidazole-resistant subtype 7 (*bottom*) isolates of *Blastocystis* in axenic cultures. Although intra-culture size variation is obvious, ST 7 isolate culture exhibits larger cell size compared to ST 4. ST 7 isolate also exhibits a homogeneous ring like appearance with a distinct large central vacuole whereas ST 4 cultures exhibit more diversity in cell shapes and vacuole formations

(Nagel et al. 2011) and high degree of genetic (Stensvold et al. 2012b) and pathobiological diversity (Fig. 5.1) within subtypes (Mirza and Tan 2009; Mirza et al. 2011a) complicate the subtype-dependent pathogenecity hypothesis.

5.2.1.3 Coinfection

Another factor undermining *Blastocystis*' status as a human pathogen is coinfections with other known intestinal pathogens (Herbinger et al. 2011). It is frequently suggested that before prescribing anti-*Blastocystis* treatment, a thorough work-up should be done to exclude other pathogens based on the assumption that co-colonization of *Blastocystis* with other known pathogens is merely a coincidence (Coyle et al. 2012; Tan et al. 2010). An underappreciated phenomenon is facilitation of intestinal colonization by opportunistic pathogens by *Blastocystis*. There are multiple reports of *Blastocystis* undermining host defenses (Dagci et al. 2002; Hussein et al. 2008; Mirza et al. 2011b, 2012; Puthia et al. 2005, 2006). The parasite increases epithelial permeability (Dagci et al. 2002; Hussein et al. 2008; Mirza et al. 2012; Puthia et al. 2006), which might assist invasion of other intestinal pathogens (Hu et al. 2008). The suppression of host immune responses by *Blastocystis* in the form of IgA degradation (Puthia et al. 2005) and epithelial nitric oxide synthase down-regulation (Mirza et al. 2011b) could also assist intestinal colonization by opportunistic pathogens (Herbinger et al. 2011).

5.2.1.4 Enteroinvasion

Two reports also suggest invasive potential of *Blastocystis* (Hu et al. 2008; Patino et al. 2008). One of these reports suggested a coinfection with *Entamoeba*, resulting in growth of *Blastocystis* in a liver abscess (Hu et al. 2008) while the other described its colonization in the peritoneum following invasive adenocarcinoma (Patino et al. 2008). The parasite has been shown to increase intestinal permeability in infection models as well as humans (Chandramathi et al. 2010a, b; Dagci et al. 2002; Hussein et al. 2008; Mirza et al. 2012; Puthia et al. 2006), however, no signs of enteroinvasion by *Blastocystis* were observed by endoscopy (Chen et al. 2003; Yakoob et al. 2004b). Overt intestinal inflammation characteristic of enteroinvasion was also absent during parasite infections (Yakoob et al. 2004b). *Blastocystis*-induced increase in intestinal permeability (Chandramathi et al. 2010a, b; Dagci et al. 2002; Hussein et al. 2008; Mirza et al. 2012; Puthia et al. 2006) makes the gut leaky resulting in diarrhea (Tan et al. 2010). It might also expose sub-epithelial tissue and extra-intestinal sites to invasion by other pathogens (Hu et al. 2008). Altogether reports of *Blastocystis* enteroinvasion are inconclusive. The parasite might gain access to extra-intestinal sites with the help of coinfections or other pathological circumstances (Hu et al. 2008; Patino et al. 2008). Such extra-intestinal colonization of *Blastocystis* could complicate management of enteroinvasive conditions such as amoebic abscess (Hu et al. 2008) and invasive carcinomas (Patino et al. 2008).

5.2.2 Irritable Bowel Syndrome

IBS is a functional disorder of gastrointestinal tract presenting with abdominal pain and altered bowel habits. Several studies implicate *Blastocystis* as a cause of irritable bowel syndrome (IBS) (Boorom et al. 2008; Stark et al. 2007; Stensvold et al. 2009c). Compared to healthy controls, a higher incidence of *Blastocystis* has been observed in IBS patients (Boorom et al. 2008; Jimenez-Gonzalez et al. 2012; Stark et al. 2007; Stensvold et al. 2009c). Specific parasite subtypes might also be associated with IBS, but the current data is inconclusive. A predominance of ST 1 was reported in one study (Yakoob et al. 2010). Another study reported presence of ST 1 and ST 3 in IBS patients from Mexico (Jimenez-Gonzalez et al. 2012), however, no difference in distribution of these strains between disease and control group was observed (Jimenez-Gonzalez et al. 2012).

5.2.2.1 Blastocystis Cysteine Proteases and IBS

There are limited studies addressing the underlying mechanism of *Blastocytis*-induced IBS. It is possible that an altered luminal environment during IBS merely favors the growth of pathogenic *Blastocystis* subtypes (Tan et al. 2010). Conversely, persistent exposure of *Blastocystis* antigens during chronic infections might induce low-grade inflammation in the intestine leading to IBS (Tan et al. 2010). Perturbations in intestinal epithelial tight junction complex with associated increase in intestinal permeability have been reported in IBS patients (Piche et al. 2009). Intestinal epithelial barrier dysfunction in IBS is attributed to activation of proteinase-activated receptors (PAR) (Bueno and Fioramonti 2008). Increase protease activity has been reported in stools of IBS patients (Gecse et al. 2008). Interestingly, there are a number of reports on increase in intestinal permeability by *Blastocystis* (Chandramathi et al. 2010a, b; Dagci et al. 2002; Hussein et al. 2008; Mirza et al. 2012; Puthia et al. 2006). In a recent study we described the role of *Blastocystis* cysteine proteases in increasing epithelial permeability by modulating tight junction complex (Mirza et al. 2012). These tight junction alterations were Rho kinase (ROCK) dependent (Mirza et al. 2012) and suspected to be due to activation of PAR by the parasite proteases (Mirza et al. 2012; Tan et al. 2010). A higher protease activity in *Blastocystis* isolates from symptomatic patients has also been reported (Mirza and Tan 2009). Furthermore, genomic data of *Blastocystis* suggests the presence of as much as 22 proteases that might be secreted by the parasite (Denoeud et al. 2011). Altogether these findings strongly support the hypothesis that *Blastocystis* proteases might play an important role in inducing IBS. Figure 5.1 illustrates the possible mechanisms for epithelial barrier disruption induced by *Blastocystis*.

5.2.3 Urticaria and Other Dermatological Disorders

A correlation between *Blastocystis* and cutaneous lesions, particularly urticaria has been increasingly reported (Hameed et al. 2011; Katsarou-Katsari et al. 2008; Vogelberg et al. 2010). Multiple case studies suggest a causal link of the parasite with acute or chronic urticaria. The literature addressing mechanisms of *Blastocystis*-induced cutaneous lesions is limited but the disorders are likely to be immune related (Tan et al. 2010). There are reports on association between delayed-pressure urticaria, angioedema, and palmoplantar pruritus with *Blastocystis* as well (Armentia et al. 1993; Tan 2008; Vogelberg et al. 2010). Cutaneous manifestations of *Blastocystis* are often associated with particular phenotype (Hameed et al. 2011;

Zuel-Fakkar et al. 2011) or subtype (Katsarou-Katsari et al. 2008; Vogelberg et al. 2010) of the parasite. A recent case report also suggests concurrence of intestinal and cutaneous symptoms during a *Blastocystis* ST2 infection (Vogelberg et al. 2010). Some studies correlate skin pathology with amoeboid form of *Blastocystis* (Hameed et al. 2011; Zuel-Fakkar et al. 2011). Eradication of *Blastocystis* from intestine results in complete resolution of skin disorders, reenforcing the role of *Blastocystis* in cutaneous pathology (Armentia et al. 1993; Pasqui et al. 2004) however recurrence of symptoms despite prolonged antimicrobial therapy has been reported (Cassano et al. 2005).

5.3 High Risk Populations

Blastocystis is considered as an opportunistic organism. Several surveys report a higher incidence of *Blastocystis* in immunocompromised individuals.

5.3.1 AIDS and HIV Patients

Blastocystis is frequently isolated from stools of HIV/AIDS patients (Tan et al. 2009). One study reports a higher incidence of *Blastocystis*-induced intestinal disorder in these patients (Idris et al. 2010) with resolution of symptoms after metronidazole treatment. A serious complication of HIV-related *Blastocystis* infections is disseminated disease, not frequently observed in otherwise healthy individuals (Kurniawan et al. 2009; Stensvold et al. 2009a). Subtype 3 appears to be the most opportunistic of *Blastocystis* subtypes since it is the most frequently isolated subtype found in HIV infected patients (Tan et al. 2009).

5.3.2 Cancer Patients

Blastocystis is one of the most common parasites found in stools of cancer patients (Taşova et al. 2000). Infections present with abdominal pain, diarrhea, and flatulence (Taşova et al. 2000). Subtype 3 and subtype 4 isolates were reported to be highly prevalent in cancer subjects (Poirier et al. 2011; Tan et al. 2009). In one report, carcinoma-induced intestinal obstruction facilitated the overt growth of *Blastocystis* (Horiki et al. 1999). Another study described growth of the parasite in abdominal fluid following bowel perforation by adenocarcinoma (Patino et al. 2008). *Blastocystis* infections have also been reported in breast and colorectal cancer patients undergoing anticancer chemotherapy (Chandramathi et al. 2012) suggesting that the parasite is capable of surviving highly cytotoxic drugs, while debilitating immunosuppressive state induced by chemotherapy (Chandramathi et al. 2012) promotes its colonization (Chandramathi et al. 2012). Altogether these reports suggest that the parasite thrives in cancer patients. Moreover *Blastocystis* colonization, particularly in patients undergoing anticancer therapy might contribute to post-chemotherapy intestinal inflammation and increased permeability (Tsuji et al. 2003).

5.3.3 Pediatric Infections

Pediatric populations are highly susceptible to *Blastocystis* infections (Londoño et al. 2009). Higher incidence of the parasite colonization is observed in immunocompetent (Londoño et al. 2009) as well as immunocompromised (Noureldin et al. 1999) minors. Children undergoing corticosteroid therapy for nephrotic syndrome were also reported to be highly susceptible to *Blastocystis* infections (Noureldin et al. 1999). Associations of low socioeconomic status (Mehraj et al. 2008) and HIV positivity (Idris et al. 2010) with *Blastocystis* infections are also observed in pediatric patients.

5.3.4 Animal Exposure

Zoonosis is frequently reported in *Blastocystis* (Stensvold et al. 2009c; Yoshikawa et al. 2009). There are several reports suggesting a higher incidence of *Blastocystis* in professionals frequently exposed to animals or animal products (Stensvold et al. 2009c). Contact with pigs, poultry, primates, and canines increases the risk of *Blastocystis* infections (Nagel et al. 2011; Navarro et al. 2008; Stensvold et al. 2009c; Yoshikawa et al. 2009). A recent study of symptomatic patients and their pets identified a concordance of parasite subtypes recovered from their stool samples suggesting a possible transmission of the parasite between them (Nagel et al. 2011).

5.4 Laboratory Diagnosis

Technical advances in recent years have made it possible to efficiently capture *Blastocystis* diversity and it should be taken into consideration during laboratory investigations (Tan et al. 2010). Information on the density, morphology, subtype, as well as antibiotic susceptibility is important for appropriate management of *Blastocystis* infections. Despite availability of tools to assess these parameters, a number of *Blastocystis* infections go unreported or ignored owing to its controversial pathogenesis.

Due to presence of diverse morphological forms, *Blastocystis* is one of the most difficult organisms to identify in stool samples. A survey comparing several European diagnostic laboratories reported discrepancies in reporting of *Blastocystis*-positive stool samples (Utzinger et al. 2010). Some reports suggest that 20–30 % of stool samples of infected individuals only contain cysts form of *Blastocystis* (Rene et al. 2009; Suresh and Smith 2004). Due to a considerably smaller size and stark morphological differences from the vacuolar form, diagnostic staff should be trained to identify cyst stage of the parasite.

Conventional diagnostic methods to identify the parasite in stools include FECT and XIVC, while tests based on serology and molecular biology are being used increasingly. *Blastocystis* antibiotic susceptibility testing is yet to be instituted in clinical diagnostics, but should be used in cases refractory to antibiotic treatment.

5.4.1 FECT

The formol ether concentration technique (FECT) is commonly used to detect parasite ova and cysts. Due to poor sensitivity of FECT, however, it is not recommended for laboratory identification of *Blastocystis* (Rene et al. 2009; Stensvold et al. 2007). The reason for the lower sensitivity might be inability of FECT to isolate subtype 3 of *Blastocystis*, a very common subtype in humans (Rene et al. 2009; Stensvold et al. 2007). Using only FECT to isolate *Blastocystis* might contribute to the underestimation of the parasite burden in epidemiologic studies (Tan et al. 2010).

5.4.2 XIVC

Trichrome-stained fecal smears following short-term xenic in vitro culture (XIVC) exhibits superior sensitivity in identification of the parasite. However, XIVC should also be used with caution. Due to difference in generation time between *Blastocystis* strains (Zierdt and Swan 1981), slower growing subtypes and mixed infections of slower- and faster growing subtypes may be missed by this technique. Another consideration for better diagnosis of *Blastocystis* infections is to take multiple stool samples over a period of time since the parasite shedding by humans is irregular (Suresh and Smith 2004).

5.4.3 Molecular Diagnostics

Emerging data suggests association of certain *Blastocystis* subtypes or strains with pathogenicity and antibiotic resistance (Stensvold et al. 2009c, 2010), highlighting the importance of subtyping in the clinical management of blastocystosis.

For accurate identification of subtypes, genotyping of *Blastocystis* by PCR methods is preferable (Stensvold et al. 2006), and should ideally be performed directly on stool samples. This could either be accomplished by sequence analysis of *Blastocystis*-specific PCR products, use of PCR primers that are specific for different parasite subtypes, or by using a subtype-specific probe based TaqMan PCR (Stensvold et al. 2012a; Yoshikawa et al. 1998).

5.4.4 Serology

There are reports of the usefulness of serological techniques in identification of *Blastocystis* infections. Some studies involving enzyme-linked immunosorbent assay for *Blastocystis* correlate antibody titers with symptoms (Hussain et al. 1997; Mahmoud and Saleh 2003). Since these techniques are quantitative and identify parasites rapidly with high sensitivity and specificity, they could be powerful tools for diagnosis of the parasite (Tan et al. 2010). Development of such assays however should take into consideration the genetic diversity of *Blastocystis*. A panel of antibodies highly specific for individual subtypes would be useful for diagnosis of *Blastocystis* (Tan et al. 2010).

5.4.5 Antibiotic Susceptibility Testing

Antibiotic susceptibility testing is not routinely performed for intestinal parasites, due to lack of knowledge regarding their resistance to conventional antibiotics. Chronic *Blastocystis* infections are often nonresponsive to conventional antibiotic regimens (Stensvold et al. 2010). Such infections are particularly difficult to manage in the absence of antibiotic susceptibility data. Classically, *Blastocystis* antibiotic susceptibility evaluation utilized serial dilution or radioisotope inclusion methods (Dunn and Boreham 1991; Yakoob et al. 2004a; Zierdt et al. 1983). These techniques were expensive, time consuming, and often injurious to health. Recent development of redox-based assays (Mirza et al. 2011a), as well as improvements in anaerobic microculture methods, has made it possible to efficiently evaluate *Blastocystis* antibiotic susceptibilities within 24 h (Mirza et al. 2011a; Upcroft and Upcroft 2001). Redox-based assays are cheap, require a small parasite density, and could be adapted on high-throughput platforms. Furthermore, these assays do not require sophisticated equipment and are ideal for use in resource-poor healthcare facilities (Mirza et al. 2011a). Despite their clinical usefulness, most of these assays are adapted for the vacuolar morphological form of the parasite (Dunn et al. 2012; Mirza et al. 2011a). It is believed that cysts, which are resistant to metronidazole (Zaman and Zaki 1996), and not the vacuolar form represent the infectious phenotype of the parasite (Tan 2008). For more accurate assessment and management of the parasite antibiotic resistance, future studies should focus on adapting antibiotic susceptibility assays for non-vacuolar morphological forms of *Blastocystis*.

5.5 Treatment

5.5.1 Current Treatment Options

Due to the controversial pathogenicity of *Blastocystis*, antibiotics are seldom prescribed for the infections caused by the parasite. Due to mild nature of the disease and self-limitation of symptoms clinicians are skeptical of antibiotic prescription for *Blastocystis* infections (Stensvold et al. 2010). Metronidazole, a position 1,5 nitroimidazole is the treatment of choice against chronic infections in which all other etiologies have been excluded (Tan et al. 2010). Multiple studies suggest the efficacy of metronidazole against *Blastocystis* (Stensvold et al. 2010). However other studies suggest that infections can be refractory to metronidazole treatment, making anti-*Blastocystis* therapy complex (Tan et al. 2010). Factors that might influence treatment outcomes include infection density, acquisition of mutations, resistance of certain developmental stages to metronidazole, and, most importantly, strain-to-strain variation in antibiotic susceptibility (Tan et al. 2010). In fact, in vitro studies have suggested a varying response of different *Blastocystis* isolates to metronidazole (Zaman and Zaki 1996; Zierdt et al. 1983). A recent study reported metronidazole resistance in subtype 7 isolates (Mirza et al. 2011a). These isolates exhibited cross resistance to tinidazole, another position 1,5 nitroimidazole commonly prescribed for parasitic infections (Mirza et al. 2011a). Although more isolates should be evaluated, these studies further highlight the significance of subtyping for effective clinical management of *Blastocystis* infections. The most commonly prescribed alternatives to metronidazole for *Blastocystis* infections are co-trimoxazole and Paromomycin (Stensvold et al. 2010). Unfortunately, resistance to these drugs has also been reported (Stensvold et al. 2010) and there is a pressing need to identify alternative treatment options for *Blastocystis* infections. Table 5.1 summarizes the clinically proven treatment options for *Blastocystis*.

5.5.2 Experimental Therapeutics

5.5.2.1 Antibiotics

The most promising alternative antibiotic for treatment of resistant *Blastocystis* isolates include furazolidone, which exhibits excellent in vitro toxicity against all clinical strains of the parasite tested so far (Mirza et al. 2011a). It is an FDA-approved nitrofuran commonly used for treatment of parasitic infections (Quiros-Buelna 1989). Other conventional antiparasitic agents highly effective against *Blastocystis* in vitro are prophylactic antimalarials mefloquine and quinacrine (Mirza et al. 2011a). Although, their clinical efficacy is not known, compared to metronidazole-sensitive strains, these agents exhibit superior antiparasitic activity

Table 5.1 Clinically evaluated antibiotics for *Blastocystis* infections

Drug	Dose	IC-50 (µM)	Clinical indication	Resistant subtypes	
				In vitro	Clinical
Metronidazole	750 mg TDS for 10 days 500 mg TDS for 10 days 1.5 g OD for 7 days	4.3–190	Intestinal and cutaneous symptoms.	ST1 and ST7 (Cysts are also resistant)	ST1, ST4 and ST6
TMP:SMz	(320 mg TMP: 1600 mg SMX) TDS for 10 days	18–22 (µg/ml)	Intestinal and cutaneous symptoms	–	ST1, ST4 and ST6
Nitazoxanide	500 mg BD for 3 days	0.6–8	Intestinal symptoms	ST4	–
Paromomycin	25 mg/kg TDS for 10 days	Not toxic in vitro	Cutaneous symptoms	–	–
Tinidazole	2 g OD for 5 days	0.3–10	–	ST4 and 7	–
Ketoconazole	200 mg OD for 7 days	10.7	–	–	–

TDS three times per day, *BD* twice daily, *OD* once daily

against metonidazole-resistant *Blastocystis* isolates (Mirza et al. 2011a) and would be ideal for persistent infections refractory to metronidazole treatment.

5.5.2.2 Probiotics

Probiotic *Saccharomyces boulardii* was reported to be useful in the treatment of *Blastocystis* infections in one study (Dinleyici et al. 2011). Pediatric patients with a 2-week long history of intestinal symptoms with stools positive for *Blastocystis* were treated with *Saccharomyces*, for 10 days. Thirty days after treatment initiation, 94.4 % of the patients became asymptomatic with resolution of stool cysts (Dinleyici et al. 2011). However, more studies are needed to establish the role of probiotics in treatment of intestinal protozoal infections (Table 5.2).

5.5.2.3 Nitric Oxide and Arginine

Cytotoxity of NO against *Blastocystis* has been reported recently (Eida et al. 2008; Mirza et al. 2011b). Parasites when exposed to nitrosative stress undergo apoptosis and/or necrosis (Eida et al. 2008; Mirza et al. 2011b). Interestingly, compared to susceptible isolate, a metronidazole-resistant isolate of *Blastocystis* was found to be more sensitive to nitrosative stress suggesting the potential usefulness of NO donors against antibiotic resistant strains (Fig. 5.2) (Mirza et al. 2011b). NO is a common innate immune response of host epithelium (Eckmann et al. 2000; Mirza et al. 2011b). It is generated by epithelial inducible nitric oxide synthase (iNOS) by using arginine as substrate (Eckmann et al. 2000). In order to facilitate its survival in gut lumen, *Blastocystis* inhibits epithelial NO generation by consuming arginine (Mirza et al. 2011b). Dietary supplementation of arginine in infected patients might be useful in restoring physiological level of luminal nitrosative stress thus limiting *Blastocystis* growth (Fig. 5.2).

5.5.2.4 Statins and Rho Kinase Inhibitors

Advances in knowledge concerning *Blastocystis*–epithelial interactions could also lead to development of novel treatment options. Observations of epithelial Rho/Rho kinase pathway modulation by the parasite led to the identification of a few unique pharmaceutical agents that prevent *Blastocystis* induced pathology (Fig. 5.2) (Mirza et al. 2012). Simvastatin is a commonly used, well-tolerated, and FDA-approved lipid-lower agent (Chen et al. 2008). One of the effects of statins is epithelial barrier protection via inhibition of Rho/Rho kinase pathway (Chen et al. 2008). A recent study suggested the role of statins in reducing the risk of *Clostridium difficile* (a Rho pathway modulating organism) intestinal infections which was independent of their lipid-lowering activity (Motzkus-Feagans et al. 2012). In our experiments we observed that *Blastocystis*-mediated increased

Table 5.2 Experimental therapeutics for *Blastocystis* infections

Drug	IC-50 (μM)	Comments
Furazolidone	0.5–1	An FDA approved nitrofuran commonly used against *Giardia* No clinical reports of resistance in *Blastocystis*
Mefloquine	1.5–5	Prophylactic antimalarial, highly toxic to metronidazole resistant *Blastocystis*
Quinacrine	3–5	Prophylactic antimalarial, highly toxic to metronidazole resistant *Blastocystis*
Ronidazole	0.3–0.5	Veterinary antiparasitic agent, equally effective against metronidazole resistant and susceptible isolates
Ornidazole	1.1–1.4	Veterinary antiparasitic agent Equally effective against metronidazole resistant and susceptible isolates
C-17	0.3–0.6	Experimental position 2,5-nitroimidazole High toxicity against *Giardia* and *Trichomonas* Exhibits cross-resistance with metronidazole in subtype 4 isolates
Nitric oxide donors		Induces apoptosis in *Blastocystis* Highly toxic against *Entamoeba* and *Giardia* Higher toxicity against metronidazole resistant isolate
Simvastatin	N/A	Commonly prescribed lipid-lowering agent Prevents parasite-induced epithelial pathology Mildly toxic to *Blastocystis*
Fasudil	N/A	Approved for treatment of stroke in Japan Prevents parasite-induced epithelial pathology
Saccharomyces boulardii	N/A	A probiotic Parasite clearance rate comparable to metronidazole

intestinal epithelial permeability is also prevented by Rho inhibition by statins (Fig. 5.2) (Mirza et al. 2012). Fasudil is a clinically useful Rho kinase inhibitor (Fig. 5.2) (Taoufiq et al. 2008). It has been shown to prevent *Plasmodium*-induced endothelial pathology (Taoufiq et al. 2008). In addition to simvastatin we also observed barrier protective effect of fasudil against *Blastocystis*-induced epithelial barrier dysfunction (Fig. 5.2) (Mirza et al. 2012). Since these agents target host cells instead of the parasite, likelihood of resistance development against these agents is limited and they could be useful adjunct or alternate treatment option against drug-resistant infections.

5.6 Discussion

Despite being one of the most common intestinal parasites reported in human stool samples, the pathogenic potential of *Blastocystis* remains controversial. The primary reasons for these controversies include lack of standardized diagnostic

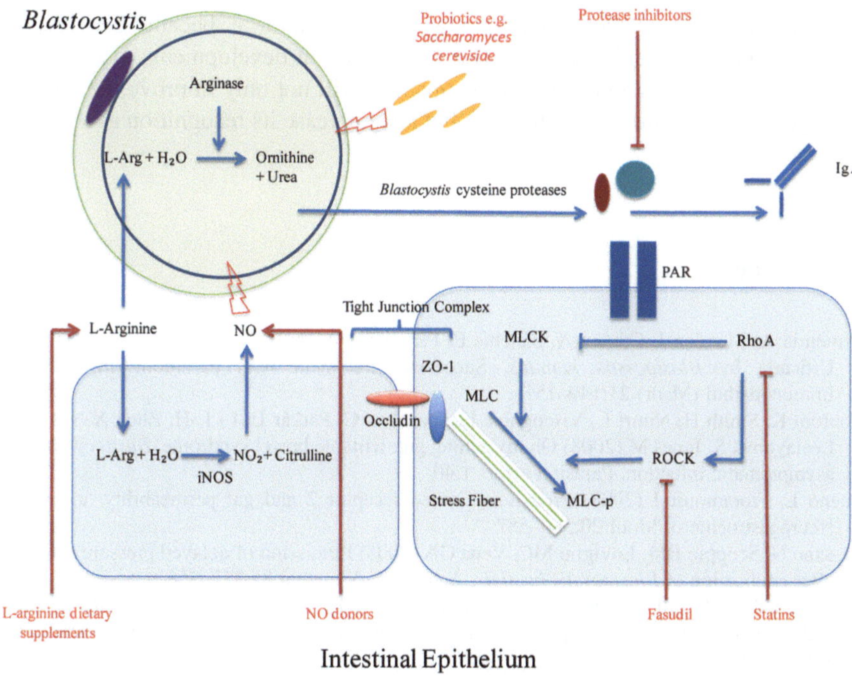

Fig. 5.2 Potential targets for therapeutic intervention. This illustration summarizes potential treatment options (*red*) and their cellular targets for *Blastocystis* infections. Drugs such as Fasudil and Simvastatin target host epithelial Rho/ROCK pathway to prevent parasite induced epithelial barrier dysfunction. NO donors mimic host innate immune generated, antiparasitic nitrosative stress to prevent parasite colonization of gut where as L-arginine supplementation could reverse parasite induced suppression of host epithelial NO generation. *Sacchromyces cerevisiae* limits parasite colonization of the gut lumen through its probiotic effect while protease inhibitors neutralize cysteine proteases, potential parasite virulence factors. *NO*, nitric oxide, *iNOS* inducible nitric oxide synthase, *ZO-1* Zonula Occludins-1, *ROCK* Rho kinase, *MLCK* myosin light chain kinase, *MLC* myosin light chain, *MLCp* phosphorylated MLC, *IgA* immunoglobulin A, *PAR* proteinase activated receptors

techniques, no consensus on treatment of the parasite, limited and variable efficacy of existing treatment regimen, and large numbers of asymptomatic carriers. Limited understanding of the complex biology of distinct *Blastocystis* phenotypes and subtypes and limited or no knowledge regarding cellular microbiology, immune evasion, and life cycle of *Blastocystis* infections further complicates the situation. All of these little-understood aspects of the parasite pathobiology are active areas of research and in the last decade a number of studies have addressed these issues to some extent. Several new diagnostic measures and treatment options are being identified along with a better understanding of the parasite's pathobiology. Due to increase in awareness of the parasite's pathogenic potential, *Blastocystis* has also been included as a microbial agent in the water sanitation and health program of the World Health Organization (WHO 2008). The clinical significance of the intestinal parasite *Giardia intestinalis* was recognized only after it became possible to clinically manage it effectively (WHO 2008). Recent developments of improved diagnostic techniques and efficient therapeutics will not only improve the clinical management of *Blastocystis* infections but also increase its recognition as a human pathogen.

References

Armentia A, Méndez J, Gómez A, Sanchís E, Fernández A, de la Fuente R, Sánchez P (1993) Urticaria by *Blastocystis hominis*. Successful treatment with paromomycin. Allergol Immunopathol (Madr) 21:149–151

Boorom K, Smith H, Nimri L, Viscogliosi E, Spanakos G, Parkar U, Li L-H, Zhou X-N, Ok U, Leelayoova S, Jones M (2008) Oh my aching gut: irritable bowel syndrome, *Blastocystis*, and asymptomatic infection. Parasit Vectors 1:40

Bueno L, Fioramonti J (2008) Protease-activated receptor 2 and gut permeability: a review. Neurogastroenterol Motil 20:580–587

Cassano N, Scoppio BM, Loviglio MC, Vena GA (2005) Remission of delayed pressure urticaria after eradication of *Blastocystis hominis*. Acta Derm Venereol 85:357–358

Chandramathi S, Suresh K, Anita ZB, Kuppusamy UR (2012) Infections of *Blastocystis hominis* and microsporidia in cancer patients: are they opportunistic? Trans R Soc Trop Med Hyg 106:267–269

Chandramathi S, Suresh K, Kuppusamy UR (2010a) Elevated levels of urinary hyaluronidase in humans infected with intestinal parasites. Ann Trop Med Parasitol 104:449–452

Chandramathi S, Suresh KG, Mahmood AA, Kuppusamy UR (2010b) Urinary hyaluronidase activity in rats infected with *Blastocystis hominis*—evidence for invasion? Parasitol Res 106:1459–1463

Chen TL, Chan CC, Chen HP, Fung CP, Lin CP, Chan WL, Liu CY (2003) Clinical characteristics and endoscopic findings associated with *Blastocystis hominis* in healthy adults. Am J Trop Med Hyg 69:213–216

Chen W, Pendyala S, Natarajan V, Garcia JG, Jacobson JR (2008) Endothelial cell barrier protection by simvastatin: GTPase regulation and NADPH oxidase inhibition. Am J Physiol Lung Cell Mol Physiol 295:L575–L583

Coyle CM, Varughese J, Weiss LM, Tanowitz HB (2012) *Blastocystis*: to treat or not to treat. Clin Infect Dis 54:105–110

Dagci H, Ustun S, Taner MS, Ersoz G, Karacasu F, Budak S (2002) Protozoon infections and intestinal permeability. Acta Trop 81:1–5

Denoeud F, Roussel M, Noel B, Wawrzyniak I, Da Silva C, Diogon M, Viscogliosi E, Brochier-Armanet C, Couloux A, Poulain J, Segurens B, Anthouard V, Texier C, Blot N, Poirier P, Ng GC, Tan KS, Artiguenave F, Jaillon O, Aury JM, Delbac F, Wincker P, Vivarès CP, El Alaoui H (2011) Genome sequence of the stramenopile *Blastocystis*, a human anaerobic parasite. Genome Biol 12:R29

Dinleyici EC, Eren M, Dogan N, Reyhanioglu S, Yargic ZA, Vandenplas Y (2011) Clinical efficacy of *Saccharomyces boulardii* or metronidazole in symptomatic children with *Blastocystis hominis* infection. Parasitol Res 108:541–545

Dogruman-Al F, Simsek Z, Boorom K, Ekici E, Sahin M, Tuncer C, Kustimur S, Altinbas A (2010) Comparison of methods for detection of *Blastocystis* infection in routinely submitted stool samples, and also in IBS/IBD Patients in Ankara, Turkey. PLoS One 5:e15484

Domínguez-Márquez MV, Guna R, Muñoz C, Gómez-Muñoz MT, Borrás R (2009) High prevalence of subtype 4 among isolates of *Blastocystis hominis* from symptomatic patients of a health district of Valencia (Spain). Parasitol Res 105:949–955

Dunn LA, Boreham PF (1991) The in-vitro activity of drugs against *Blastocystis hominis*. J Antimicrob Chemother 27:507–516

Dunn LA, Tan KS, Vanelle P, Juspin T, Crozet MD, Terme T, Upcroft P, Upcroft JA (2012) Development of metronidazole-resistant lines of *Blastocystis* sp. Parasitol Res 111(1):441–450

Eckmann L, Laurent F, Langford TD, Hetsko ML, Smith JR, Kagnoff MF, Gillin FD (2000) Nitric oxide production by human intestinal epithelial cells and competition for arginine as potential determinants of host defense against the lumen-dwelling pathogen *Giardia lamblia*. J Immunol 164:1478–1487

Eida OM, Hussein EM, Eida AM, El-Moamly AA, Salem AM (2008) Evaluation of the nitric oxide activity against *Blastocystis hominis* in vitro and in vivo. J Egypt Soc Parasitol 38:521–536

Eroglu F, Genc A, Elgun G, Koltas IS (2009) Identification of *Blastocystis hominis* isolates from asymptomatic and symptomatic patients by PCR. Parasitol Res 105:1589–1592

Gecse K, Róka R, Ferrier L, Leveque M, Eutamene H, Cartier C, Ait-Belgnaoui A, Rosztóczy A, Izbéki F, Fioramonti J, Wittmann T, Bueno L (2008) Increased faecal serine protease activity in diarrhoeic IBS patients: a colonic lumenal factor impairing colonic permeability and sensitivity. Gut 57:591–599

Hameed DM, Hassanin OM, Zuel-Fakkar NM (2011) Association of *Blastocystis hominis* genetic subtypes with urticaria. Parasitol Res 108:553–560

Herbinger KH, Fleischmann E, Weber C, Perona P, Löscher T, Bretzel G (2011) Epidemiological, clinical, and diagnostic data on intestinal infections with *Entamoeba histolytica* and Entamoeba dispar among returning travelers. Infection 39:527–535

Horiki N, Kaneda Y, Maruyama M, Fujita Y, Tachibana H (1999) Intestinal blockage by carcinoma and *Blastocystis hominis* infection. Am J Trop Med Hyg 60:400–402

Hu KC, Lin CC, Wang TE, Liu CY, Chen MJ, Chang WH (2008) Amoebic liver abscess or is it? Gut 57(627):683

Hussain R, Jaferi W, Zuberi S, Baqai R, Abrar N, Ahmed A, Zaman V (1997) Significantly increased IgG2 subclass antibody levels to *Blastocystis hominis* in patients with irritable bowel syndrome. Am J Trop Med Hyg 56:301–306

Hussein EM, Hussein AM, Eida MM, Atwa MM (2008) Pathophysiological variability of different genotypes of human *Blastocystis hominis* Egyptian isolates in experimentally infected rats. Parasitol Res 102:853–860

Idris NS, Dwipoerwantoro PG, Kurniawan A, Said M (2010) Intestinal parasitic infection of immunocompromised children with diarrhoea: clinical profile and therapeutic response. J Infect Dev Ctries 4:309–317

Jimenez-Gonzalez DE, Martinez-Flores WA, Reyes-Gordillo J, Ramirez-Miranda ME, Arroyo-Escalante S, Romero-Valdovinos M, Stark D, Souza-Saldivar V, Martinez-Hernandez F, Flisser A, Olivo-Diaz A, Maravilla P (2012) *Blastocystis* infection is associated with irritable bowel syndrome in a Mexican patient population. Parasitol Res 110:1269–1275

Katsarou-Katsari A, Vassalos CM, Tzanetou K, Spanakos G, Papadopoulou C, Vakalis N (2008) Acute urticaria associated with amoeboid forms of *Blastocystis* sp. subtype 3. Acta Derm Venereol 88:80–81

Kaya S, Cetin E, Aridoğan B, Arikan S, Demirci M (2007) Pathogenicity of *Blastocystis hominis*, a clinical reevaluation. Turkiye Parazitol Derg 31:184–187

Kurniawan A, Karyadi T, Dwintasari SW, Sari IP, Yunihastuti E, Djauzi S, Smith HV (2009) Intestinal parasitic infections in HIV/AIDS patients presenting with diarrhoea in Jakarta, Indonesia. Trans R Soc Trop Med Hyg 103:892–898

Londoño AL, Mejía S, Gómez-Marín JE (2009) Prevalence and risk factors associated with intestinal parasitism in preschool children from the urban area of Calarcá, Colombia. Rev Salud Publica (Bogota) 11:72–81

Mahmoud MS, Saleh WA (2003) Secretory and humoral antibody responses to *Blastocystis hominis* in symptomatic and asymptomatic human infections. J Egypt Soc Parasitol 33:13–30

Mehraj V, Hatcher J, Akhtar S, Rafique G, Beg MA (2008) Prevalence and factors associated with intestinal parasitic infection among children in an urban slum of Karachi. PLoS ONE 3:e3680

Mirza H, Tan KS (2009) *Blastocystis* exhibits inter- and intra-subtype variation in cysteine protease activity. Parasitol Res 104:355–361

Mirza H, Teo JD, Upcroft J, Tan KS (2011a) A rapid, high-throughput viability assay for *Blastocystis* spp. reveals metronidazole resistance and extensive subtype-dependent variations in drug susceptibilities. Antimicrob Agents Chemother 55:637–648

Mirza H, Wu Z, Kidwai F, Tan KS (2011b) A metronidazole-resistant isolate of *Blastocystis* spp. is susceptible to nitric oxide and downregulates intestinal epithelial inducible nitric oxide synthase by a novel parasite survival mechanism. Infect Immun 79:5019–5026

Mirza H, Wu Z, Teo JD, Tan KS (2012) Statin-pleiotropy prevents rho kinase-mediated intestinal epithelial barrier compromise induced by *Blastocystis* cysteine proteases. Cell Microbiol 14(9):1474–1484

Moghaddam DD, Ghadirian E, Azami M (2005) *Blastocystis hominis* and the evaluation of efficacy of metronidazole and trimethoprim/sulfamethoxazole. Parasitol Res 96:273–275

Motzkus-Feagans CA, Pakyz A, Polk R, Gambassi G, Lapane KL (2012) Statin use and the risk of *Clostridium difficile* in academic medical centres. Gut. 2012 Mar 22. [Epub ahead of print]

Nagel R, Cuttell L, Stensvold CR, Mills PC, Bielefeldt-Ohmann H, Traub RJ (2011) *Blastocystis* subtypes in symptomatic and asymptomatic family members and pets and response to therapy. Intern Med J

Navarro C, Domínguez-Márquez M, Garijo-Toledo M, Vega-García S, Fernández-Barredo S, Pérez-Gracia M, García A, Borrás R, Gómez-Muñoz M (2008) High prevalence of *Blastocystis* sp. in pigs reared under intensive growing systems: frequency of ribotypes and associated risk factors. Vet Parasitol 153:347–358

Noureldin MS, Shaltout AA, El Hamshary EM, Ali ME (1999) Opportunistic intestinal protozoal infections in immunocompromised children. J Egypt Soc Parasitol 29:951–961

Pasqui AL, Savini E, Saletti M, Guzzo C, Puccetti L, Auteri A (2004) Chronic urticaria and *Blastocystis hominis* infection: a case report. Eur Rev Med Pharmacol Sci 8:117–120

Patino WD, Cavuoti D, Banerjee SK, Swartz K, Ashfaq R, Gokaslan T (2008) Cytologic diagnosis of *Blastocystis hominis* in peritoneal fluid: a case report. Acta Cytol 52:718–720

Piche T, Barbara G, Aubert P, Bruley des Varannes S, Dainese R, Nano JL, Cremon C, Stanghellini V, De Giorgio R, Galmiche JP, Neunlist M (2009) Impaired intestinal barrier integrity in the colon of patients with irritable bowel syndrome: involvement of soluble mediators. Gut 58:196–201

Poirier P, Wawrzyniak I, Albert A, El Alaoui H, Delbac F, Livrelli V (2011) Development and evaluation of a real-time PCR assay for detection and quantification of *Blastocystis* parasites in human stool samples: prospective study of patients with hematological malignancies. J Clin Microbiol 49:975–983

Puthia M, Vaithilingam A, Lu J, Tan K (2005) Degradation of human secretory immunoglobulin A by *Blastocystis*. Parasitol Res 97:386–389

Puthia MK, Sio SW, Lu J, Tan KS (2006) Blastocystis ratti induces contact-independent apoptosis, F-actin rearrangement, and barrier function disruption in IEC-6 cells. Infect Immun 74:4114–4123

Quiros-Buelna E (1989) Furazolidone and metronidazole for treatment of giardiasis in children. Scand J Gastroenterol Suppl 169:65–69

Rene BA, Stensvold CR, Badsberg JH, Nielsen HV (2009) Subtype analysis of *Blastocystis* isolates from *Blastocystis* cyst excreting patients. Am J Trop Med Hyg 80:588–592

Salvi PF, Virgilio E, Bocchetti T, Costa G, Pascarella G, Balducci G (2012) *Blastocystis hominis* and recurrent megacolon: a causative or fortuitous association? Am Surg 78:198–199

Stark D, van Hal S, Marriott D, Ellis J, Harkness J (2007) Irritable bowel syndrome: a review on the role of intestinal protozoa and the importance of their detection and diagnosis. Int J Parasitol 37:11–20

Stensvold CR, Ahmed UN, Andersen LO, Nielsen HV (2012a) Development and evaluation of a genus-specific, probe-based, internal process controlled real-time PCR assay for sensitive and specific detection of Blastocystis. J Clin Microbiol 50(6):1847–1851

Stensvold CR, Alfellani M, Clark CG (2012b) Levels of genetic diversity vary dramatically between *Blastocystis* subtypes. Infect Genet Evol 12:263–273

Stensvold CR, Alfellani MA, Nørskov-Lauritsen S, Prip K, Victory EL, Maddox C, Nielsen HV, Clark CG (2009a) Subtype distribution of *Blastocystis* isolates from synanthropic and zoo animals and identification of a new subtype. Int J Parasitol 39:473–479

Stensvold CR, Arendrup MC, Jespersgaard C, Mølbak K, Nielsen HV (2007) Detecting *Blastocystis* using parasitologic and DNA-based methods: a comparative study. Diagn Microbiol Infect Dis 59:303–307

Stensvold CR, Arendrup MC, Nielsen HV, Bada A, Thorsen S (2008) Symptomatic infection with *Blastocystis* sp. subtype 8 successfully treated with trimethoprim-sulfamethoxazole. Ann Trop Med Parasitol 102:271–274

Stensvold CR, Arendrup MC, Nielsen HV, Mølbak K (2009b) *Blastocystis*—an enigmatic parasite. Ugeskr Laeger 171:2388–2390

Stensvold CR, Lewis HC, Hammerum AM, Porsbo LJ, Nielsen SS, Olsen KE, Arendrup MC, Nielsen HV, Mølbak K (2009c) *Blastocystis*: unravelling potential risk factors and clinical significance of a common but neglected parasite. Epidemiol Infect 137:1655–1663

Stensvold CR, Smith HV, Nagel R, Olsen KE, Traub RJ (2010) Eradication of *Blastocystis* carriage with antimicrobials: reality or delusion? J Clin Gastroenterol 44:85–90

Stensvold R, Brillowska-Dabrowska A, Nielsen HV, Arendrup MC (2006) Detection of *Blastocystis hominis* in unpreserved stool specimens by using polymerase chain reaction. J Parasitol 92:1081–1087

Suresh K, Smith H (2004) Comparison of methods for detecting *Blastocystis hominis*. Eur J Clin Microbiol Infect Dis 23:509–511

Tan KS (2008) New insights on classification, identification, and clinical relevance of *Blastocystis* spp. Clin Microbiol Rev 21:639–665

Tan KS, Mirza H, Teo JD, Wu B, Macary PA (2010) Current views on the clinical relevance of *Blastocystis* spp. Curr Infect Dis Rep 12:28–35

Tan TC, Ong SC, Suresh KG (2009) Genetic variability of *Blastocystis* sp. isolates obtained from cancer and HIV/AIDS patients. Parasitol Res 105:1283–1286

Taoufiq Z, Gay F, Balvanyos J, Ciceron L, Tefit M, Lechat P, Mazier D (2008) Rho kinase inhibition in severe malaria: thwarting parasite-induced collateral damage to endothelia. J Infect Dis 197:1062–1073

Taşova Y, Sahin B, Koltaş S, Paydaş S (2000) Clinical significance and frequency of *Blastocystis hominis* in Turkish patients with hematological malignancy. Acta Med Okayama 54:133–136

Tsuji E, Hiki N, Nomura S, Fukushima R, Kojima J, Ogawa T, Mafune K, Mimura Y, Kaminishi M (2003) Simultaneous onset of acute inflammatory response, sepsis-like symptoms and intestinal mucosal injury after cancer chemotherapy. Int J Cancer 107:303–308

Upcroft JA, Upcroft P (2001) Drug susceptibility testing of anaerobic protozoa. Antimicrob Agents Chemother 45:1810–1814

Utzinger J, Botero-Kleiven S, Castelli F, Chiodini PL, Edwards H, Köhler N, Gulletta M, Lebbad M, Manser M, Matthys B, N'Goran EK, Tannich E, Vounatsou P, Marti H (2010) Microscopic diagnosis of sodium acetate-acetic acid-formalin-fixed stool samples for helminths and intestinal protozoa: a comparison among European reference laboratories. Clin Microbiol Infect 16:267–273

Vogelberg C, Stensvold CR, Monecke S, Ditzen A, Stopsack K, Heinrich-Gräfe U, Pöhlmann C (2010) *Blastocystis* sp. subtype 2 detection during recurrence of gastrointestinal and urticarial symptoms. Parasitol Int 59:469–471

WHO (2008) Guideline for drinking water quality. Incorporated first and second addenda, Geneva, p 514

Yakoob J, Jafri W, Beg MA, Abbas Z, Naz S, Islam M, Khan R (2010) Irritable bowel syndrome: is it associated with genotypes of *Blastocystis hominis*. Parasitol Res 106:1033–1038

Yakoob J, Jafri W, Jafri N, Islam M, Asim Beg M (2004a) In vitro susceptibility of *Blastocystis hominis* isolated from patients with irritable bowel syndrome. Br J Biomed Sci 61:75–77

Yakoob J, Jafri W, Jafri N, Khan R, Islam M, Beg MA, Zaman V (2004b) Irritable bowel syndrome: in search of an etiology: role of *Blastocystis hominis*. Am J Trop Med Hyg 70:383–385

Yoshikawa H, Nagano I, Wu Z, Yap EH, Singh M, Takahashi Y (1998) Genomic polymorphism among *Blastocystis hominis* strains and development of subtype-specific diagnostic primers. Mol Cell Probes 12:153–159

Yoshikawa H, Wu Z, Pandey K, Pandey BD, Sherchand JB, Yanagi T, Kanbara H (2009) Molecular characterization of *Blastocystis* isolates from children and rhesus monkeys in Kathmandu, Nepal. Vet Parasitol 160:295–300

Zaman V, Zaki M (1996) Resistance of *Blastocystis hominis* cysts to metronidazole. Trop Med Int Health 1:677–678

Zierdt C, Swan J (1981) Generation time and growth rate of the human intestinal parasite *Blastocystis hominis*. J Protozool 28:483–485

Zierdt CH, Swan JC, Hosseini J (1983) In vitro response of *Blastocystis hominis* to antiprotozoal drugs. J Protozool 30:332–334

Zuel-Fakkar NM, Abdel Hameed DM, Hassanin OM (2011) Study of *Blastocystis hominis* isolates in urticaria: a case-control study. Clin Exp Dermatol 36:908–910

Chapter 6
Molecular Approaches on Systematic Position, Genetic Polymorphism, and Classification of *Blastocystis* Isolates from Humans and Animals

Hisao Yoshikawa

Abstract *Blastocystis* was first reported as a harmless yeast found in human stool samples in 1912. This organism has been classified into the subphylum Sporozoa and then reclassified into the subphylum Sarcodina based on the morphological and physiological characteristics. Molecular sequencing study on the SSU rRNA gene of a human *Blastocystis* isolate showed that *Blastocystis* is not monophyletic with yeasts (*Saccharomyces*), fungi (*Neurospora*), sarcodines (*Naegleria*, *Acanthamoeba*, and *Dictyostelium*), or sporozoans (*Sarcocystis* and *Toxoplasma*) based on the partial sequence of the gene. Phylogenetic studies inferred with the entire sequence of the SSU rRNA gene of a human and a guinea pig isolates of *Blastocystis* revealed that the organism is a member of the stramenopiles. Furthermore, accumulation of the molecular data of human and animal *Blastocystis* isolates showed not only extensive genetic diversity among *Blastocystis* isolates from the same hosts but also genetic similarities were observed among the isolates from different hosts. Therefore, a consensus terminology of *Blastocystis* sp. is recently proposed for the isolates from mammalian and avian hosts including humans and it classifies all isolates into nine subtypes. Recently, however, several additional novel subtypes have been reported based on the phylogenetic studies of the partial sequences of the SSU rRNA gene. Therefore, the genus *Blastocystis* may be the most genetically polymorphic organism that parasitizes a wide range of animals.

6.1 Introduction

Blastocystis hominis was originally isolated from humans and reported as a harmless yeast by Brumpt (1912). The taxonomic classification was first challenged by morphological and physiological characteristics in cultures and *Blastocystis*

H. Yoshikawa (✉)
Department of Biological Sciences, Nara Women's University, Nara, Japan
e-mail: h.yoshikawa@cc.nara-wu.ac.jp

H. Mehlhorn et al. (eds.), *Blastocystis: Pathogen or Passenger?*,
Parasitology Research Monographs 4,
DOI 10.1007/978-3-642-32738-4_6,

showed many protozoan characteristics rather than yeast (Zierdt et al. 1967). Physiologically, *Blastocystis* is strictly anaerobic and growth of the organisms is supported in some intestinal protozoan media, while no growth in fungal media. The optimal pH and temperature for growth are 7.0–8.0 and 37 °C, while no growth under than pH 6.0 and 33 °C, respectively. Morphologically, *Blastocystis* lacks a cell wall, contains nucleus, smooth and rough endoplasmic reticulum, Golgi complex, and mitochondria-like organelle. Subsequently, this organism was proposed as a sporozoan and then reclassified as a sarcodine by Zierdt and colleagues based on the more detail morphological characteristics in in vitro cultures (Zierdt and Tan 1976; Zierdt 1991). However, morphological features of this parasite do not fit to any other protozoa. Therefore, molecular studies are essential to estimate or classify systematic position among a wide range of the phylum protozoa.

6.2 Molecular Approach on Phylogenetic Affinity

The fist molecular systematic approach was undertaken by using the parasite's small subunit rRNA (SSU rRNA) gene sequences with those from different eukaryotes (Johnson et al. 1989). Based on the partial sequences, it was shown that *Blastocystis* is not monophyletic with the yeasts (*Sacchromyces*), sarcodines (*Naegleria, Acanthamoeba*, and *Dictyostelium*) or sporozoan (*Sarcocystis* and *Toxoplasma*). It was concluded that close relationships to the yeasts or the apicomplexas/sporozoans are not supported. A later study (Silberman et al., 1996) revealed entire sequence of the SSU rRNA gene of two *Blastocystis* isolates from a human and a guinea pig and it showed that *Blastocystis* could be a member of the heterogeneous group of stramenopiles. Stramenopiles are a diverse heterogeneous group of unicellular and multicellular algae and protists, including heterotropic and photosynthetic representatives (Patterson 1989). This diverse group includes brown algae, diatoms, oomycetes, and several other lineages. The closest relative of *Blastocystis* in the stramenopiles is *Proteromonas lacertae*, a commensal flagellate in the hindgut of lizards and amphibia (Leipe et al. 1996). However, unlike other stramenopiles, no flagella and tubular hairs have ever been observed in *Blastocystis*.

Another molecular approach has also been employed with the amino acid sequence of a highly conserved gene, elongation factor 1α (EF-1α). By using a maximum likelihood method of protein phylogeny, it was concluded that *Blastocystis* is not close to fungi and the divergence of the parasite is earlier than those of *Trypanosoma*, *Euglena*, *Dictyostelium*, and other higher eukaryotes (Nakamura et al. 1996). However, it was not possible to resolve the positions of *Blastocystis, Entamoeba, Plasmodium*, and *Tetrahymena* due to the low bootstrap probabilities. The subsequent study showed that, based on the genetic distance between homologous genes, *Blastocystis* from humans and animals diverged within the same group and the isolates all belonged to the same genus (Ho et al. 2000). The data also suggested an earlier emergence of *Blastocystis* among eukaryotes and a close relative of *Entamoeba histolytica*. Interestingly, these molecular phylogenetic studies of *Blastocystis* SSU rRNA and EF-1 gene sequences are apparently disparate conclusions on taxonomic or phylogenetic affiliations. Therefore, molecular approaches on systematic position of *Blastocystis* was conducted with multiple conserved genes, SSU rRNA, cytosolic-type 70-kDa heat shock protein, traslation elongation factor 2, and the non-catalytic 'B' subunit of vacuolar ATPase of *Blastocystis* with other stramenopiles (Arisue et al. 2002). This study clearly demonstrated that *Blastocystis* is a stramenopile and also alveolates are the closest relatives of *Blastocystis*/stramenopiles. More recently, phylogenetic analysis based on the conserved 5.8S rRNA gene with 12 stramenopiles, 2 fungi, 3 algae, and 3 alveolates showed *Blastocystis* positioned within stranenopiles, with *Proteromonas lacertae* as its closest relative (Hoevers and Snowden 2005).

6.3 Genetic Polymorphism and Molecular Classification

A large number of molecular studies have revealed that *Blastocystis* isolates from humans and animals appear to be extensively polymorphic in genetics (Yoshikawa et al. 1996, 1998, 2000, 2004a, b, c; Böhm-Gloning et al. 1997; Clark 1997; Hoevers et al. 2000; Snowden et al. 2000; Kaneda et al. 2001; Abe et al. 2003a, b, c; Abe 2004; Noël et al. 2003, 2005). In the earlier studies, random amplified polymorphic DNA (RAPD) analysis and restriction fragment length polymorphism (RFLP) analysis of the small subunit ribosomal RNA (SSU rRNA) genes have been applied to reveal genetic diversity among *Blastocystis* populations from humans to animals (Yoshikawa et al. 1996, 1998; Böhm-Gloning et al. 1997; Clark 1997; Hoevers et al. 2000; Snowden et al., 2000; Kaneda et al. 2001; Abe et al. 2003a, b). Namely, genetic variations within human and animal *Blastocystis* isolates were observed by RAPD analysis (Yoshikawa et al. 1996, 1998). An isolate from a chicken showed similar RAPD profiles with human isolates from USA and Japan, but not with five isolates from Singapore. This result indicates the genomic polymorphism among *Blastocystis* isolates from humans and animals and also a possibility of zoonotic potential of the parasite. In addition, subtype classification has been proposed for different genotypes among human *Blastocystis* isolates revealed by RAPD profiles and then developed subtype-specific PCR primers from the unique sequences of the RAPD products (Yoshikawa et al. 1998, 2000, 2003). Subsequently, subtype classification is generally accepted for consensus terminology for the genus *Blastocystis* (Stensvold et al. 2007). Conversely, Clark (1997) showed the sequence diversity among 30 human *Blastocystis* isolates using RFLP analysis of the amplified SSU rRNA genes and proposed seven ribodeme classification (riboprinting) among the isolates. This study showed extensive genetic diversity, over 7 % divergence between ribodemes 1 and 2 of human *Blastocystis* isolates. Surprisingly, this is approximately four times the genetic distance between the homologous genes of *Entamoeba histolytica* and *E. disper*. However, recent accumulation of RFLP analysis and sequencing of the SSU rRNA gene of human and animal isolates, it became apparent that RFLP approach is limited for use for classification of *Blastocystis* isolates. Therefore, genetic polymorphism among various human and animal *Blastocystis* isolates has been investigated based on the sequencing of the SSU rRNA gene.

Accumulated data of the SSU rRNA gene sequences of various *Blastocystis* isolates from humans and animals indicate that most of the phylogenetic clade show a mixed population of *Blastocystis* isolates from both humans and animals inferred with the SSU rRNA gene (Arisue et al. 2003; Noël et al. 2003; Yoshikawa et al. 2004c). These results strongly suggest that proposing of various new species names based mainly on host origins are problematic and cause confusion among *Blastocystis* isolates from various hosts (Yoshikawa et al. 2004c). Therefore, a consensus terminology of *Blastocystis* sp. is recently proposed for the isolates from mammalian and avian animal hosts including humans and it classifies all isolates into 9 subtypes (STs) based on the phylogeny inferred with SSU rRNA gene (Stensvold et al. 2007) (Table 6.1). The reason that *Blastocystis* sp. rather than *Blastocystis hominis* was proposed is because some reptilian and amphibian species seems to fall within the range of variation by the mammalian and avian clades (Yoshikawa et al. 2004c; Noël et al. 2005). If the reptilian or amphibian isolates are distinct species, all of the nine clades (nine subtypes) cannot be *Blastocystis hominis*. However, a most recent phylogenetic study inferred with the homologous gene including more wide range of isolates from vertebrate to invertebrate hosts has apparently shown that most isolates from poikilothermal animal hosts (i.e., reptiles, amphibians, and insects) are positioned into the basal clades among the genus of *Blastocystis*, suggesting that the isolates from poikilothermal animal hosts are different from homoiothermal animal hosts (i.e., mammals and birds) (Yoshikawa et al. 2007).

Based on the subtype classification, the majority of human *Blastocystis* belong to subtype 3, while *Blastocystis* isolated from several populations such as Iran, Malaysia, Pakistan, Philippines, and Thailand belong to subtype 1 (Table 6.1). However, the sample size of the latter group is relatively small in number (12–45 isolates) and discrepancy results were observed in the same countries in Pakistan and Thailand. Therefore more detail research should be accomplished in these countries to confirm which subtype is the dominant among the human *Blastocystis* populations. Moreover, recent findings of human infections with *Blastocystis* sp. STs 5, 8 and 9 (Yoshikawa et al. 2004a, b, c; Yan et al. 2007; Stensvold et al. 2008) clearly show that human is specific host capable of being infected by any *Blastocystis* subtypes. Recently, there have been reported additional novel STs from ST10 to ST13 based mainly on partial sequence (about 300–1,000 bp) of the SSU rRNA gene (Li et al., 2007a; Parkar et al. 2010; Stensvold et al. 2009b). However, there is no guarantee that phylogenetic tree inferred with partial sequence will fit to the tree inferred with the entire sequence of the same gene. Since subtype classification for *Blastocystis* sp. was originally inferred with the entire sequence of the SSU rRNA gene, full sequence of the SSU rRNA gene should deposit in GenBank as a reference sequence data for the novel subtypes.

Table 6.1 *Blastocystis* subtypes among human populations in various countries

Country	No. of isolates	Subtype (ST) 1	2	3	4	5	6	7	8	9	Unknown	Mixed ST	References
Bangladesh	26	2	0	24	0	0	0	0	0	0	0	0	Yoshikawa et al. (2004b)
China	35	13	2	14	0	0	0	0	0	0	1	5	Yan et al. (2006)
	78	16	1	55	1	0	0	0	0	0	3	2	Li et al. (2007b)
	192	47	9	116	1	0	1	0	0	0	8	10	Li et al. (2007a)
Denmark	99[a]	20	15	39	16	0	1	0	1	0	0	7	Rene et al. (2009)
	206	37	52	50	35	0	0	6	0	1	0	25	Stensvold et al. (2009a)
Egypt	44	8	4	24	8	0	0	0	0	0	0	0	Hussein et al. (2008)
	20	3	4	12	0	0	0	0	0	0	0	1	Souppart et al. (2010)
	100	15	0	39	0	0	23	13	0	0	0	10	Fouad et al. (2011)
Philippines	12	10[b]	0	0	0	0	0	0	0	0	2	0	Rivera and Tan (2005)
France	40	8	4	20	4	0	0	1	0	0	0	3	Souppart et al. (2009)
Germany	166	35	1	110	12	0	0	0	0	0	0	8	Böhm-Gloning et al. (1997)
	12	3	2	5	0	2	0	0	0	0	0	0	Yoshikawa et al. (2004b)
Greece	45	9	6	27	1	0	1	1	0	0	0	0	Menounos et al. (2008)
	51	7	5	32	1	0	1	5	0	0	0	0	Vassalos et al. (2010)
Iran	45	20[b]	4	16	0	0	2	3	0	0	0	0	Motazedian et al. (2008)
Italy	32	2	6	16	6	0	0	1	1	0	0	0	Meloni et al. (2011))
Japan	19	4	6	5	4	0	0	0	0	0	0	0	Kaneda et al. (2001)
	50	4	0	26	2	0	11	5	0	2	0	0	Yoshikawa et al. (2004b)
Malaysia	17	9[b]	1	7	0	0	0	0	0	0	0	0	Tan et al. (2008)
Nepal	20	4	4	12	0	0	0	0	0	0	0	0	Yoshikawa et al. (2009)
	63	2	0	0	12	0	0	0	0	0	8	41	Lee et al. (2012)
Pakistan	10	2	0	7	0	0	1	0	0	0	0	0	Yoshikawa et al. (2004b)
	179	87[b]	10	49	8	7	6	10	0	0	0	2	Yakoob et al. (2010)
Singapore	13	2	4	7	0	0	0	0	0	0	0	0	Wong et al. (2008)
Spain	51	1	2	0	48	0	0	0	0	0	0	0	Domínguez-Márquez et al. (2009)

(continued)

Table 6.1 (continued)

Country	No. of isolates	Subtype (ST)											References
		1	2	3	4	5	6	7	8	9	Unknown	Mixed ST	
Sweden	63	10	9	30	13	0	0	1	0	0	0	0	Forsell et al. (2012)
Thailand	153	7	0	138	2	0	0	0	0	0	0	6	Thathaisong et al. (2003)
	68	53[b]	15	0	0	0	0	0	0	0	0	0	Leelayoova et al. (2008)
Turkey	92	17	20	51	0	0	0	0	0	0	0	4	Dogruman-Al et al. (2008)
	87	8	12	66	1	0	0	0	0	0	0	0	Özyurt et al. (2008)
	32	0	3	29	0	0	0	0	0	0	0	0	Eroglu et al. (2009)
	19	0	8	10	0	0	0	0	0	0	0	1	Dogruman-Al et al. (2009a)
	66	10	9	38	0	0	0	0	0	0	0	9	Dogruman-Al et al. (2009b)
	50	8	5	32	0	0	0	0	0	0	0	5	Yoshikawa et al. (2011)
UK	30	3	1	22	4	0	0	0	0	0	0	0	Clark (1997)
	56	3	9	22	21	0	0	0	0	0	0	1	Scicluna et al. (2006)
US	7	1	0	6	0	0	0	0	0	0	0	0	Jones et al. (2009)
	5	1	1	1	1	0	0	0	1	0	0	0	Whipps et al. (2010)
Total	2297	487	225	1135	180	9	47	46	3	3	22	139	

[a] A total number of samples were obtained from 74 patients.

[b] The subtype 1 is the major subtype in these countries.

6.4 Conclusion

Ever since the first description of a harmless yeast *Blastocystis hominis* in humans in 1912, this eukaryotic microorganism has been classified into various subphylum based on morphological and physiological characteristics. However, recent accumulation of the molecular data of several conserved genes indicates that the genus *Blastocystis* is a member of the stramenopiles. Furthermore, an extensive genetic diversity has been recognized among *Blastocystis* isolates from humans and animals indicating that this parasite may be one of the most indistinguishable unicellular parasite composing from heterogeneous groups of different species. Because, it has been reported that the 629 positions from the university core region of the SSU rRNA gene, that should be very conservative, the maximum difference among the *Blastocystis* isolates was shown to be 3.6 %, which was comparable with the differences between *Saccharomyces cerevisiae* and *Pneumocystis carinii* (3.2 %) and between *Zea mays* and *Chlorella kessleri* (3.8 %) (Arisue et al. 2003). In addition, this protozoan parasite may not be host specific and many animal species are potential sources for reservoirs of zoonotic isolates. Further molecular research may facilitate species designation for this parasite and pathogenic potential of the specific genotype or species.

References

Abe N, Wu Z, Yoshikawa H (2003a) Zoonotic genotypes of *Blastocystis hominis* detected in cattle and pigs by PCR with diagnostic primers and restriction fragment length polymorphism analysis of the small subunit ribosomal RNA gene. Parasitol Res 90:124–128

Abe N, Wu Z, Yoshikawa H (2003b) Molecular characterization of *Blastocystis* isolates from birds by PCR with diagnostic primers and restriction fragment length polymorphism analysis of the small subunit ribosomal RNA gene. Parasitol Res 89:393–396

Abe N, Wu Z, Yoshikawa H (2003c) Molecular characterization of *Blastocystis* isolates from primates. Vet Parasitol 113:321–325

Abe N (2004) Molecular and phylogenetic analysis of *Blastocystis* isolates from various hosts. Vet Parasitol 120:235–242

Arisue N, Hashimoto T, Yoshikawa H, Nakamura Y, Nakamura G, Nakamura F, Yano TA, Hasegawa M (2002) Phylogenetic position of *Blastocystis hominis* and of stramenopiles inferred from multiple molecular sequence data. J Eukaryot Microbiol 49:42–53

Arisue N, Hashimoto T, Yoshikawa H (2003) Sequence heterogeneity of the small subunit ribosomal RNA genes among *Blastocystis* isolates. Parasitology 126:1–9

Böhm-Gloning B, Knobloch J, Walderich B (1997) Five subgroups of *Blastocystis hominis* from symptomatic and asymptomatic patients revealed by restriction site analysis of PCR-amplified 16S-like rDNA. Trop Med Int Health 2:771–778

Brumpt E (1912) *Blastocystis hominis* n sp. et formes voisines. Bull Soc Pathol Exot 5:725–730

Clark CG (1997) Extensive genetic diversity in *Blastocystis hominis*. Mol Biochem Parasitol 87:79–83

Dogruman-Al F, Dagci H, Yoshikawa H, Kurt Ö, Demirel M (2008) A possible link between subtype 2 and asymptomatic infections of *Blastocystis hominis*. Parasitol Res 103:685–689

Dogruman-Al F, Kustimur S, Yoshikawa H, Tuncer C, Simsek Z, Tanyuksel M, Araz E, Boorom K (2009a) *Blastocystis* subtypes in irritable bowel syndrome and inflammatory bowel diseases in Ankara, Turkey. Mem Inst Oswaldo Cruz (Rio de Janeiro) 104:724–727

Dogruman-Al F, Yoshikawa H, Kustimur S, Balaban N (2009b) PCR-based subtyping of *Blastocystis* isolates from symptomatic and asymptomatic individuals in a major hospital in Ankara, Turkey. Parasitol Res 106:263–268

Domínguez-Márquez MV, Guna R, Muñoz C, Gómez-Muñoz T, Borrás R (2009) High prevalence of subtype 4 among isolates of *Blastocystis hominis* from symptomatic patients of a health district of Valencia (Spain). Parasitol Res 105:949–955

Eroglu F, Genc A, Elgun G, Koltas IS (2009) Identification of *Blastocystis hominis* isolates from asymptomatic and symptomatic patients by PCR. Parasitol Res 105:1589–1592

Forsell J, Granlund M, Stensvold CR, Clark GC, Evengard B (2012) Subtype analysis of *Blastocystis* isolates in Swedish patients. Eur J Clin Microbiol Infect Dis 31:1689–1696

Fouad SA, Basyoni MMA, Fahmy RA, Kobaisi MH (2011) The pathogenic role of different *Blastocystis hominis* genotypes isolated from patients with irritable bowel syndrome. Arab J Gastroenterol 12:194–200

Ho LC, Armugam A, Jeyaseelan K, Yap EH, Singh M (2000) *Blastocystis* elongation factor-1α: genomic organization, taxonomy and phylogenetic relationships. Parasitology 121:135–144

Hoevers JD, Snowden KF (2005) Analysis of the ITS region and partial ssu and lsu rRNA genes of *Blastocystis* and *Proteromonas lacertae*. Parasitology 131:187–196

Hoevers J, Holman P, Logan K, Hommel M, Ashford R, Snowden K (2000) Restriction-fragment-length polymorphism analysis of small-subunit rRNA genes of *Blastocystis hominis* isolates from geographically diverse human hosts. Parasitol Res 86:57–61

Hussein EM, Hussein AM, Eida MM, Atwa MM (2008) Pathophysiological variability of different genotypes of human *Blastocystis hominis* Egyptian isolates in experimentally infected rats. Parasitol Res 102:853–860

Jones MS, Whipps CM, Ganac RD, Hudson NR, Boroom K (2009) Association of *Blastocystis* subtype 3 and 1 with patients from an Oregon community presenting with chronic gastrointestinal illness. Parasitol Res 104:341–345

Johnson AM, Thanou A, Boreham PFL, Baverstock PR (1989) *Blastocystis hominis*: phylogenetic affinities determined by rRNA sequence comparison. Exp Parasitol 68:283–288

Kaneda Y, Horiki N, Cheng X-J, Fujita Y, Maruyama M, Tchibana H (2001) Ribodemes of *Blastocystis hominis* isolated in Japan. Am J Trop Med Hyg 65:393–396

Lee IL, Tan TC, Tan PC, Nanthiney DR, Biraj MK, Surendra KM, Suresh KG (2012) Predominance of *Blastocystis* sp. subtype 4 in rural communities, Nepal. Parasitol Res 110:1553–1562

Leelayoova S, Siripattanapipong S, Thathaisong U, Naaglor T, Taamasri P, Piyaraj P, Mungthin M (2008) Drinking water: a possible source of *Blastocystis* spp. subtype 1 infection in schoolchildren of a rural community in central Thailand. Am J Trop Med Hyg 79:401–406

Leipe DD, Tong SM, Goggin CL, Slemenda SB, Pieniazek NJ, Sogin ML (1996) 16S-like rDNA sequences from *Developayella elegans*, *Labyrinthuloides haliotidis*, and *Proteromonas lacertae* confirm that the stramenopiles are a primarily heterotrophic group. Eur J Protistol 32:449–458

Li LH, Zhang X-P, Lv S, Zhang L, Yoshikawa H, Wu Z, Steinmann P, Utzinger J, Tong X-M, Chen S-H, Zhou X-N (2007a) Cross-sectional surveys and subtype classification of human *Blastocystis* isolates from four epidemiological setting in China. Parasitol Res 102:83–90

Li LH, Zhou X-N, Du Z-W, Wang X-Z, Wang L-B, Jiang J-Y, Yoshikawa H, Steinmann P, Utzinger J, Wu Z, Chen J-X, Chen S-H, Zhang L (2007b) Molecular epidemiology of human *Blastocystis* in a village in Yunnan province, China. Parasitol Int 56:281–286

Meloni D, Sanciu G, Poirier P, El Alaoui H, Chabé M, Delhaes L, Dei-Cas E, Delbac F, Fiori PL, Di Cave D, Viscogliosi E (2011) Molecular subtyping of *Blastocystis* sp. isolates from symptomatic patients in Italy. Parasitol Res 109:613–619

Menounos PG, Spanakos G, Tegos N, Vassalos CM, Papadopoulou C, Vakalis NC (2008) Direct detection of *Blastocystis* sp. in human faecal samples and subtype assignment using single strand conformational polymorphism and sequencing. Mol Cell Probes 22:24–29

Motazedian H, Ghasemi H, Sadjjadi SM (2008) Genomic diversity of *Blastocystis hominis* from patients in southern Iran. Ann Trop Med Parasitol 102:85–88

Nakamura Y, Hashimoto T, Yoshikawa H, Kamaishi T, Nakamura F, Okamoto K, Hasegawa M (1996) Phylogenetic position of *Blastocystis hominis* that contains cytochrome-free mitochondria, inferred from the protein phylogeny of elongation factor 1α. Mol Biochem Parasitol 77:241–245

Noël C, Peyronnet C, Gerbod D, Edgcomb VP, Delgado-Viscogliosi P, Sogin ML, Capron M, Viscogliosi E, Zenner L (2003) Phylogenetic analysis of *Blastocystis* isolates from different hosts based on the comparison of small-subunit rRNA gene sequences. Mol Biochem Parasitol 126:119–123

Noël C, Dufernez F, Gerbod D, Edgcomb VP, Delgado-Viscogliosi P, Ho L-C, Singh M, Wintjens R, Sogin ML, Capron M, Pierce R, Zenner L, Viscogliosi E (2005) Molecular phylogenies of *Blastocystis* isolates from different hosts: implications for genetic diversity, identification of species, and zoonosis. J Clin Microbiol 43:348–355

Özyurt M, Kurt Ö, Mølbak K, Nielsen HV, Haznedaroglu T, Stensvold CR (2008) Molecular epidemiology of *Blastocystis* infections in Turkey. Parasitol Int 57:300–306

Parkar U, Traub RJ, Vitali S, Elliot A, Levecke B, Robertson I, Geurden T, Steele J, Drake B, Thompson RCA (2010) Molecular characterization of *Blastocystis* isolates from zoo animals and their animal-keepers. Vet Parasitol 169:8–17

Patterson DJ (1989) Stramenopiles: chromophytes from a protistan perspective. In: Green JC, Leadbeater BSC, Diver WL (eds) The chromophyte algae: problems and perspectives. Clarendon, Oxford, pp 357–379

Rene BA, Stensvold CR, Badsberg JH, Nielsen HV (2009) Subtype analysis of *Blastocystis* isolates from *Blastocystis* cyst excreting patients. Am J Trop Med Hyg 80:588–592

Rivera WL, Tan MA (2005) Molecular characterization of *Blastocystis* isolates in the Philippines by riboprinting. Parasitol Res 96:253–257

Scicluna SM, Tawari B, Clark CG (2006) DNA barcoding of *Blastocystis*. Protistology 157:77–85

Silberman JD, Sogin ML, Leipe DD, Clark CG (1996) Human parasite finds taxonomic home. Nature 380:398

Snowden K, Logan K, Blozinski C, Hoevers J, Holman P (2000) Restriction-fragment-length polymorphism analysis of small-subunit rRNA genes of *Blastocystis* isolates from animal hosts. Parasitol Res 86:62–66

Souppart L, Sanciu G, Cian A, Wawrzyniak I, Delbac F, Capron M, Dei-Cas E, Boorom K, Delhaes L, Viscogliosi E (2009) Molecular epidemiology of human *Blastocystis* isolates in France. Parasitol Res 105:413–421

Souppart L, Moussa H, Cian A, Sanciu G, Poirier P, El Alaoui H, Delbac F, Boorom K, Delhaes L, Dei-Cas E, Viscogliosi E (2010) Subtype analysis of *Blastocystis* isolates from symptomatic patients in Egypt. Parasitol Res 106:505–511

Stensvold CR, Suresh GK, Tan KSW, Thompson RCA, Traub RJ, Viscogliosi E, Yoshikawa H, Clark CG (2007) Terminology for *Blastocystis* subtypes—a consensus. Trends Parasitol 23:93–96

Stensvold CR, Arendrup MC, Nielsen HV, Bada A, Thorsen S (2008) Symptomatic infection with *Blastocystis* sp. subtype 8 successfully treated with trimethoprim-sulfamethoxazole. Ann Trop Med Parasitol 102:271–274

Stensvold CR, Lewis HC, Hammerum AM, Porsbo LJ, Nielsen SS, Olsen KEP, Arendrup MC, Nielsen HV, Mølbak K (2009a) *Blastocystis*: unravelling potential risk factors and clinical significance of a common but neglected parasite. Epidemiol Infect 137:1655–1663

Stensvold CR, Alfellani MA, Nørskov-Lauritsen S, Prip K, Victory EL, Maddox C, Nielsen HV, Clark CG (2009b) Subtype distribution of *Blastocystis* isolates from synanthropic and zoo animals and identification of a new subtype. Int J Parasitol 39:473–479

Tan TC, Suresh KG, Smith HV (2008) Phenotypic and genotypic characterization of *Blastocystis hominis* isolates implicates subtype 3 as a subtype with pathogenic potential. Parasitol Res 104:85–93

Thathaisong U, Worapong J, Mungthin M, Tan-Ariya P, Viputtigul K, Sudatis A, Noonai A, Leelayoova S (2003) *Blastocystis* isolates from a pig and a horse are closely related to *Blastocystis hominis*. J Clin Microbiol 41:967–975

Vassalos CM, Spanakos G, Vassalou E, Papadopoulou C, Vakalis N (2010) Differences in clinical significance and morphologic features of *Blastocystis* sp subtype 3. Am J Clin Pathol 133:251–258

Whipps CM, Boorom K, Bermudez LE, Kent ML (2010) Molecular characterization of *Blastocystis* species in Oregon identifies multiple subtypes. Parasitol Res 106:827–832

Wong KHS, Ng GC, Lin RTP, Yoshikawa H, Taylor MB, Tan KSW (2008) Predominance of subtype 3 among *Blastocystis* isolates from a major hospital in Singapore. Parasitol Res 102:663–670

Yakoob J, Jafri W, Beg MA, Abbas Z, Naz S, Islam M, Khan R (2010) Irritable bowel syndrome: is it associated with genotypes of *Blastocystis hominis*. Parasitol Res 106:1033–1038

Yan Y, Su S, Lai R, Liao H, Ye J, Li X, Luo X, Chen G (2006) Genetic variability of *Blastocystis hominis* isolates in China. Parasitol Res 99:597–601

Yan Y, Su S, Ye J, Lai X, Lai R, Liao H, Chen G, Zhang R, Hou Z, Luo X (2007) *Blastocystis* sp. subtype 5: a possibly zoonotic genotype. Parasitol Res 101:1527–1532

Yoshikawa H, Nagano I, Yap EH, Singh M, Takahashi Y (1996) DNA polymorphism revealed by arbitrary primers polymerase chain reaction among *Blastocystis* strains isolated from humans, a chicken, and a reptile. J Eukaryot Microbiol 43:127–130

Yoshikawa H, Nagano I, Wu Z, Yap EH, Singh M, Takahashi Y (1998) Genomic polymorphism among *Blastocystis hominis* strains and development of subtype-specific diagnostic primers. Mol Cell Probes 12:153–159

Yoshikawa H, Abe N, Iwasawa M, Kitano S, Nagano I, Wu Z, Takahashi Y (2000) Genomic analysis of *Blastocystis hominis* strains isolated from two long-term health care facilities. J Clin Microbiol 38:1324–1330

Yoshikawa H, Wu Z, Nagano I, Takahashi Y (2003) Molecular comparative studies among *Blastocystis* isolates obtained from humans and animals. J Parasitol 89:585–594

Yoshikawa H, Abe N, Wu Z (2004a) PCR-based identification of zoonotic isolates of *Blastocystis* from mammals and birds. Microbiology 150:1147–1151

Yoshikawa H, Wu Z, Kimata I, Iseki M, Ali IKMD, Hossain MB, Zaman V, Haque R, Takahashi Y (2004b) Polymerase chain reaction-based genotype classification among *Blastocystis hominis* populations isolated from different countries. Parasitol Res 92:22–29

Yoshikawa H, Morimoto K, Wu Z, Singh M, Hashimoto T (2004c) Problems in speciation in the genus *Blastocystis*. Trend Parasitol 20:251–255

Yoshikawa H, Wu Z, Howe J, Hashimoto T, Geok-Choo N, Tan KSW (2007) Ultrastructural and phylogenetic studies on *Blastocystis* isolates from cockroaches. J Eukaryot Microbiol 54:33–37

Yoshikawa H, Wu Z, Pandy K, Pandy BD, Sherchand JB, Yanagi T, Kanbara H (2009) Molecular characterization of *Blastocystis* isolates from children and rhesus monkeys in Kathmandu, Nepal. Vet Parasitol 160:295–300

Yoshikawa H, Dogruman-Al F, Turk S, Kustimur S, Balaban N, Sultan N (2011) Evaluation of DNA extraction kits for molecular diagnosis of human *Blastocystis* subtypes from fecal samples. Parasitol Res 109:1045–1050

Zierdt CH, Rude WS, Bull BS (1967) Protozoan characteristics of *Blastocystis hominis*. Am J Clin Pathol 48:495–501

Zierdt CH, Tan HK (1976) Ultrastructure and light microscope appearance of *Blastocystis hominis* in a patient with enteric diseases. Z Parasitenkd 50:277–283

Zierdt CH (1991) *Blastocystis hominis*—past and future. Clin Microbiol Rev 4:61–79

Chapter 7
What Do We Know About *Blastocystis* Analyzing Research Studies with Statistical Methods

Kenneth Boorom

Abstract *Blastocystis* is the most commonly identified parasitic infection in stool samples from healthcare-seeking patients in many regions of the world. A collection of close to 1,000 studies exists on *Blastocystis*, with substantial contributions from researchers in the Middle East, Asia/Pacific Rim, Europe, and North America. In many cases, disagreements about *Blastocystis* appear to originate from difficulties in determining how decisions should be made from the existing database of studies. Although more studies will certainly be available in the future, it is possible to apply objective screening techniques to studies which can be expected to identify conclusions that are unlikely to be reversed. However, the conclusions identified with this technique are not necessarily the same conclusions described as certain by medical professionals. A better understanding of the meaning of statistical data obtained from populations would help reduce contradictory studies, and examples are provided of situations under which a pathogen like *Giardia intestinalis* will appear more often in healthy individuals. The role of *Blastocystis* in irritable bowel syndrome (IBS) is examined, and formulas presented to help understand the costs of blastocystosis, given research studies identifying the cost of IBS.

Keywords *Blastocystis* • Pathogenicity • Irritable bowel syndrome • Diarrheal diseases

---

Electronic supplementary material: The online version of this article (doi: 10.1007/978-3-642-32738-4) contains supplementary material, which is available to authorized users.

K. Boorom (✉)
Blastocystis Research Foundation, 5060 SW Philomath Blvd, #202, Corvallis, OR 97333, USA
e-mail: kboorom@bhomcenter.org

H. Mehlhorn et al. (eds.), *Blastocystis: Pathogen or Passenger?*,
Parasitology Research Monographs 4,
DOI 10.1007/978-3-642-32738-4_7, © Springer-Verlag Berlin Heidelberg 2012

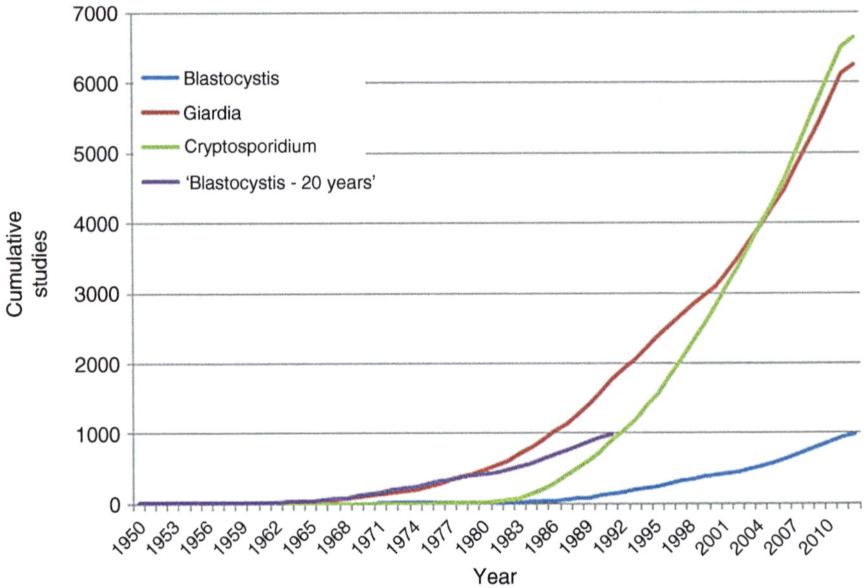

Fig. 7.1 Cumulative NIH indexed studies for three common intestinal protists. In terms of the total number of studies, *Blastocystis* spp. lags *Giardia intestinalis* by 30 years, and *Cryptosporidium* spp. by 20 years. Most of the controversy concerning *Blastocystis* infection occurred between 1986 and 1993, when there were very few studies on the infection available

7.1 Introduction

As of May 2012, the US National Institute of Health (NIH) Pubmed medical research database shows that 935 studies referring to *Blastocystis* have been cataloged. With close to 1,000 papers on the subject, it is likely that such a collection contains useful information, misleading information, as well as studies which contradict each other's findings. Given that this situation exists for *Blastocystis* (as well as most other microbes of clinical interest), how can we extract information that is needed from that database for clinical and research purposes? How can we use those studies to make decisions? Can we make decisions based on this quantity of studies?

Certainly, there are fewer studies on *Blastocystis* than either *Giardia intestinalis* or *Cryptosporidium* spp., each of which has roughly 6,000 studies, or six times as many papers as *Blastocystis* (Figs. 7.1 and 7.2). Each year, about 350 new *G. intestinalis* papers are added to Pubmed's database. Each year a similar quantity of *Cryptosporidium* sp., research is added to that database, while 50–60 new studies on *Blastocystis* are added annually (Fig. 7.2). Physicians and researchers did make decisions concerning *G. intestinalis* and *Cryptosporidium* spp. when the collection of studies for those microbes consisted of only 1,000 papers, in the 1980s and

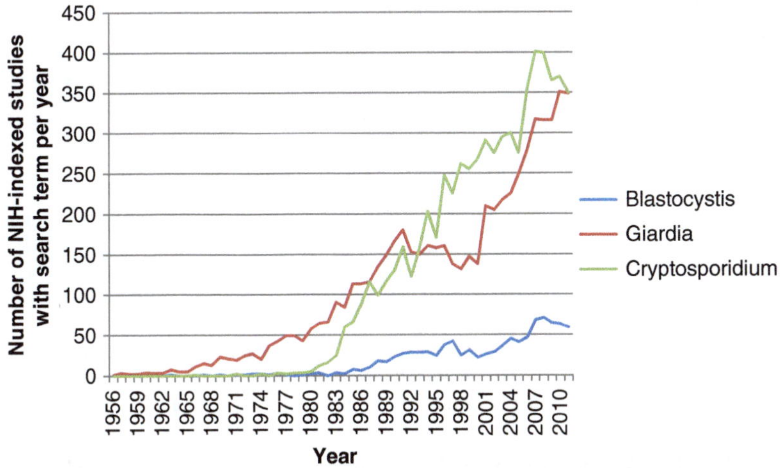

Fig. 7.2 Number of new NIH indexed studies published each year for three common intestinal Protists. The rate of publication for *G. intestinalis* and *Cryptosporidium* spp. is about 5–7 times higher than that for *Blastocystis*

Table 7.1 Comparison of the number of studies, the number of new studies, and NIH grant approval for three gastrointestinal protozoa

Metric	Selection criteria	*Blastocystis*	Giardia (Giardia : *Blastocystis*)	Cryptosporidium
Total studies published	Microbe name in title, abstract, or keyword	935	6,023	6,311
Total studies with microbe name in title	Microbe name in title	574	2,871	3,594
Average new studies published annually, 2009–2011	Microbe name in title, abstract, or keyword	56	310	325
Number of US National Institute of Health grants awarded 2000–2011[a]	Microbe name in title, abstract, or keyword	0	366	693

The total number of studies and the number of new studies for *G. intestinalis* and *Cryptosporidium* spp. outnumber *Blastocystis* by a factor of 5–7, and the finding is relatively invariant depending on the two selection criteria used for defining a microbe study *Blastocystis*.
[a]Search performed with NIH Project Reporter tool, 5/5/2012. http://projectreporter.nih.gov/reporter.cfm

1990s. There is little evidence that decisions made based on that smaller set of studies were egregiously wrong, or resulted in widespread harm or misdiagnosis of patients. As such, it is reasonable to suggest that we can use a database consisting of 1,000 studies to make responsible decisions concerning *Blastocystis* (Table 7.1).

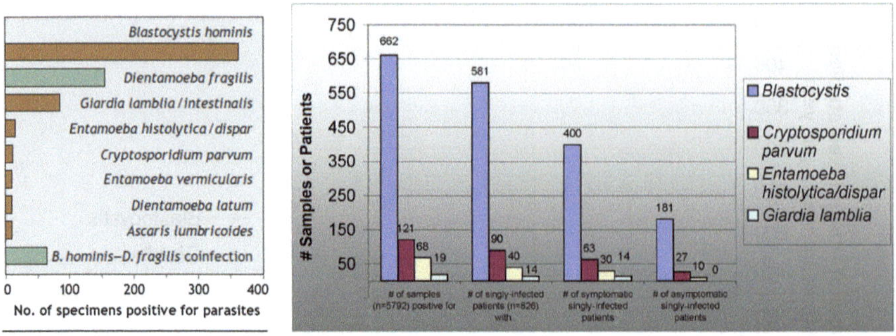

Fig. 7.3 *Blastocystis* infection dominates stool samples submitted by patients to clinical laboratories in most developed countries. Results from stool examination of patients at a regional Canadian laboratory in 2005 and a US study of samples collected from all 50 states in 2000

7.2 Should We Make Any Decisions About *Blastocystis*?

Before examining ways of making decisions, it may be worth acknowledging two contrarian arguments:

1. Sufficient studies do not exist to make a decision with confidence, so no decision should be made.
2. The existence of conflicting studies on *Blastocystis* pathogenicity proves that the microbe could be only marginally pathogenic, and as such, it warrants no attention.

In examining the first argument, it should be noted that in developed countries, symptomatic *Blastocystis* mono-infections now outnumber *G. intestinalis*, *Cryptosporidium* spp., and *Entamoeba histolytica* mono-infections combined in parasitological samples from symptomatic patients (Fig. 7.3). That is, a physician ordering such examinations for patients with gastrointestinal illness will need to make some kind of decision as to whether to prescribe antimicrobial treatment, pursue further testing, or to diagnose the patient with irritable bowel syndrome (IBS). Because physicians cannot delay decisions about patients for several years, physicians will make decisions in these situations using the available information. As such, the choice presented to us is not between a decision and no decision, but between a decision based on current scientific studies, or a decision based on cultural beliefs, notes from an old parasitology course, or other less desirable sources. Even if we could prohibit clinical decision making for *Blastocystis*, it would still be necessary to make decisions about whether research should be performed.

In addressing the second argument, it is noted that evidence of *Blastocystis* nonpathogenicity has come almost exclusively from statistical analysis of population studies, and the statistical methods used in those studies are not guaranteed to prove pathogenicity, only correlation. A pathogen can become uncorrelated with illness when other factors, such as innate or acquired immunity, enter into the population and confound the study results. Additionally, as noted later in the chapter, the presence of other microbes that produce symptoms equivalent to blastocystosis will confound a population study if those patients so-infected are also *Blastocystis*-negative.

One interesting example of this comes from studies relating infection with *Helicobacter pylori* and gastric cancer. Researchers used two different immunoassays to assess *H. pylori* infection, namely a CagA assay and a whole cell *H. pylori* assay were used to investigate cancer risk. The population consisted of 181 gastric cancer patients and 193 cancer-free controls. Taken individually, a positive finding for either of these immunoassays showed no statistically significant correlation with gastric cancer originating in the cardia, and no correlation with gastric cancer originating elsewhere ($p = 0.07, 0.11$). However, when the patient populations for both cancers were combined, a statistically significant relationship was formed with the whole cell response ($p = 0.03$). Furthermore, when the two immunoassays were combined, a statistically significant relationship was found between a positive antigenic finding, and both types of gastric cancer individually, and both types combined ($p = 0.02, p = 0.01, p = 0.003$, respectively).

In the case of *H. pylori* and gastric cancer, we can see that a serious illness can be correlated with a microbe, even though some study constructions do not show a correlation. This addresses the second argument, mainly that if all studies do not show a microbe is correlated with an illness, then the microbe must be a mild pathogen. It may be worth noting that in the case of *Blastocystis* studies, many of the most frequently cited studies, especially the ones that produced conflicting information, did not specify how patients were divided into symptomatic and asymptomatic groups with enough detail to allow us to replicate them. That is, we do not know how the symptoms typically found in this patient population, such as diarrhea, vomiting, nausea, fatigue, abdominal pain, and constipation would translate into an assignment to the symptomatic group or asymptomatic group. This is particularly problematic for enteric protozoal infections, as individuals with diarrhea commonly constitute only a third of the symptomatic group (Qadri et al. 1989; US EPA 1979).

7.3 What Kind of Decisions Need to Be Made Concerning *Blastocystis*?

Although the discussion concerning *Blastocystis* is often portrayed as "pathogenic versus nonpathogenic," the actual decisions that concern the microbe can span a much larger range. Examples of decisions that face institutions and individuals include:

1. If *Blastocystis* is found in a patient who is symptomatic, what actions should be taken?

   (a) Should additional testing be performed?
   (b) If additional testing is negative, should treatment be attempted?
   (c) Should the patient be informed about the infection?
   (d) If the patient is informed, what additional information should be provided?

2. If researchers studying psychosomatic illness perform a study on patients with "irritable bowel syndrome," should those researchers required to test patients for parasitological infections first?
   If patients in such a study are positive for *Blastocystis*, should they be informed of the infection?
   Is it acceptable to require neuro-gastroenterologists (NGs) to inform IBS patients of *Blastocystis* infection if those researchers disagree with the emphasis placed on the germ theory in identifying gastrointestinal illness, for example as noted in introduction to the Rome III Diagnostic Criteria for Functional Gastrointestinal Disorders (Drossman 2006).
   If NG's must test IBS patients for infectious diseases, should microbiologists be required to test diarrheal patients for things of interest to NG's, such as early life exposure to emotional stress or traumatic instances of material separation? (Drossman 2006).
3. Should this country have a *Blastocystis* research effort?
4. Should lettuce or spinach contaminated with *Blastocystis* cysts be fed to humans?
5. If a food service worker at an elementary school develops chronic diarrhea, and tests positive for *Blastocystis*, should that worker be informed of the infection? If a particular physician opposes classification of *Blastocystis* as a pathogen, is it acceptable for that physician to diagnose such an individual with IBS and refer them to psychotherapy?
6. If some physicians believe that the food service worker should be informed of the infection, but other physicians indicate that no information should be provided, is that an acceptable situation?
7. Can an employer require a food service employee to be treated for Blastocystis infection if they have been shown to have infected others through food handling practices?
   What if the employee's physician opposes classification of Blastocystis as a pathogen?

7.4 What Is a Metastudy?

Scientists often perform studies on populations of patients, but how do we perform a study on a population of research papers? Unfortunately, there are few standards or consensus decisions for examining a collection of studies and drawing a conclusion.[1] The Cochrane Reviews may be the most familiar example of a

[1] In 2010, BRF submitted an early version of our metastudy on *Blastocystis* research to a journal, and that paper consisted mostly of a list of studies finding *Blastocystis* pathogenic or nonpathogenic. A reviewer rejected the paper on the basis that scientific findings are largely the result of researchers repeating studies until they obtain the result they want, and as such surveying papers or tabulating study results was of no value. While this is an interesting idea, it creates a kind of philosophical crisis, in that we would then be unable to use any scientific study result, if such results are fabrications of researchers.

methodological approach to decision making, but historically these have been limited to decision making about the efficacies of treatment, rather than assessment of clinical significance of microbial infections.[2]

One approach to decision making involves performing a "metastudy" and using statistical methods similar to those used in patient population studies. A metastudy analyzes research findings in the same way that a clinical study analyzes patients:

1. An initial selection criteria is applied to the database of studies. For example, a search term like *Blastocystis* might be provided to a search engine like the NIH's Pubmed database.
2. From the studies returned in step #1, further selection criteria are applied to eliminate studies that do not address the question being examined. For example, a metastudy on antibiotic resistance might discard studies which did not examine this particular phenomenon.
3. One or more study parameters are quantified and examined for correlation. For example, a study that investigates whether physical therapy is valuable in the rehabilitation of hamstring injuries might record the number of weeks each study noted until return to normal function, along with the type of physical therapy (if any) provided.

A metastudy differs from a review in that the studies being examined are selected using a process which can be applied by other researchers in such a way that the findings would be repeatable. Reviews can largely be the product of a specific researcher, and as such, can be skewed in favor of the author's particular view, and are therefore not repeatable. The lack of repeatability can be seen by comparing different reviews on the same subject.

7.5 The First *Blastocystis* Metastudy

By 2008, the NIH's Pubmed server database had a total of 670 studies on *Blastocystis*, but nobody had ever performed a metastudy on this group of studies. Moreover, it appeared that some public agencies were engaging in a very selective process when identifying which studies were deemed relevant concerning *Blastocystis* significance. At BRF, we organized a diverse group of 11 *Blastocystis* researchers from 9 countries to examine this collection of research and also to examine the relationship between *Blastocystis* and IBS (Boorom et al. 2008). In terms of researcher participation and number of countries represented, this study may be the largest published on either *Blastocystis* or IBS (Table 7.2).

[2] We contacted the Cochrane Reviews group about this possibility, but they indicated that such a study would likely be a low priority, because *Blastocystis* infection was not viewed as germane to global world health goals.

Table 7.2 Co-authors of a metastudy of *Blastocystis* research published in 2010 in BMC Parasites and Vectors

No.	Name	Affiliation	Country
1	Kenneth Boorom	Blastocystis Research Foundation, Corvallis, OR	USA
2	Huw Smith	Scottish Parasite Diagnostic Laboratory Glasgow	UK
3	Laila Nimri	US Center for Disease Control and Jordan University of Science and Technology Atlanta, GA	Jordan
4	Eric Visclgliosi	Pasteur Institute, Lille, France	France
5	Gregory Spanakos	National School of Public Health Athens	Greece
6	Unaiza Parkar	WHO Collaborating Centre for the Molecular Epidemiology of Parasitic Infections, School of Veterinary and Biomedical Sciences, Murdoch University	Australia
7	Lan-Hua Li	Department of Preventative Medicine Weifang Medical University	China
8	Xiao-Nong Zhou	China National Institute of Parasitic Diseases	China
9	Ulgen Ok	Department of Parasitology Celal Bayer University	Turkey
10	Saovanee Leelayoova	Phramongkutklao College of Medicine Bangkok	Thailand
11	Morris Jones	US Air Force Travis Air Force Base, CA	USA

We invited researchers from nine countries to participate in order to minimize the impact of cultural beliefs on the study's findings

For each of the *Blastocystis* studies examined, we collected data about sample size, study approach (animal studies, population studies, treatment studies), the year of the study, etc. We also recorded the country where the study was performed. We then sought to determine if any of those factors were related to findings of pathogenicity.

At the time, a line of investigation had suggested that certain *Blastocystis* subtypes may be pathogenic or nonpathogenic, and this was responsible for disagreement among researchers. We were, expecting to see some kind of geographical clustering, but as happens sometimes in such studies, our actual findings were very different and potentially more valuable:

1. We analyzed 680 studies, 102 of which provided a finding concerning pathogenicity or nonpathogenicity. We excluded review studies, and letters to the editor, surveys of enteric protozoa which did not report symptoms, etc. (Fig. 7.4).
2. Within the 102 studies, we found 16 studies identifying *Blastocystis* as nonpathogenic, and 86 studies identifying it as pathogenic (Table 7.3).
3. All (16/16) studies identifying *Blastocystis* as nonpathogenic were conducted on individuals from more affluent countries (Europe, US, and Australia). Half of the 16 studies (8/16) identifying *Blastocystis* as nonpathogenic were performed in the USA before 1994. When studies performed outside of the USA, or after 1994 were considered, 93 % (79/85) of such studies identified *Blastocystis* as pathogenic.

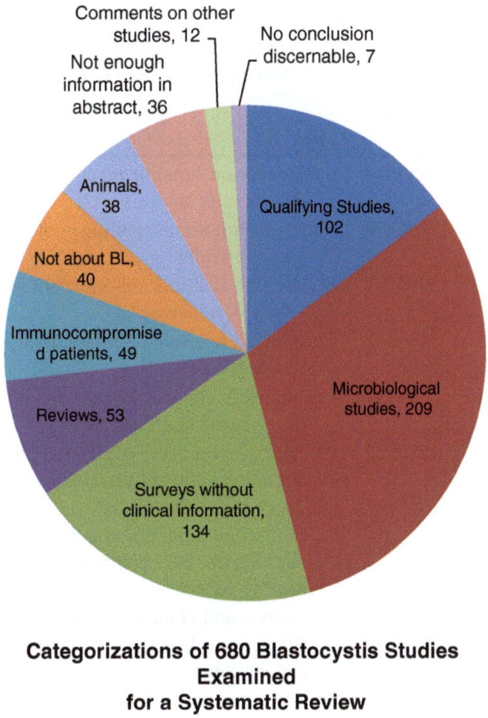

Fig. 7.4 Analysis of NIH-Indexed *Blastocystis* Studies as of January 2008. Qualifying studies were those related to infection in immunocompetent individuals where the researcher reported a finding concerning pathogenicity based on a scientific investigation

4. Overall, 44 % (8/18) of the studies performed in North America before 1994 identified *Blastocystis* as nonpathogenic, while 93 % (79/85) of the studies performed after 1994 or outside of North America identified *Blastocystis* as pathogenic.
5. In almost all studies (15/16) concluding that *Blastocystis* was nonpathogenic, the author had identified a specific property of *Blastocystis* infection thought to be incompatible with pathogenicity. For example, the presence of symptomatic and asymptomatic individuals in the same family was thought to be impossible if *Blastocystis* were a pathogen (Senay and MacPherson 1990).

We mapped researcher findings for those studies conducted in the Western Hemisphere, because most of the disagreement appeared to originate in this region (Fig. 7.5). While some clustering existed, we could not explain the divergence of opinion in terms of the spread of different *Blastocystis* subtypes, primarily because of item #5 above. Mainly, researcher finding varied not because *Blastocystis* varied, but because researchers themselves differed.

The most fascinating example of this involved one study published in 1993 by a pair of California physicians (Dr. Edward K. Markell and Dr. Udkow) that noted *Blastocystis* presence was uncorrelated with symptoms in patients from Oakland, California (Udkow and Markell 1993). This paper became one of the five works selected by the Center for Disease Control for that group's *Blastocystis* Fact Sheet, which was distributed from the CDC's web site for over 10 years, from 2000 to 2010 (CDC 2000) (Table 7.4)

Table 7.3 In a survey of 680 *Blastocystis* studies published before January 2008, we identified 16 that found *Blastocystis* to be nonpathogenic, and 86 that found it to be pathogenic. Studies finding *Blastocystis* to be nonpathogenic are shown in this table.

No.	Pubmed ID	Title	Author	*n* Subjects	Year	Location
1	2889924	Lack of serum immune response to *Blastocystis* hominis	Chen		1987	USA
2	3766850	*Blastocystis* hominis: pathogen or fellow traveler?	Markell	148	1986	USA
3	3055191	*Blastocystis* hominis: an organism in search of a disease	Miller		1988	USA
4	8463612	*Blastocystis* hominis in inflammatory bowel disease	Nagler	12	1993	USA
5	2596457	Questionable clinical significance of *Blastocystis* hominis infection	Sun	6,262	1989	USA
6	2229995	Frequency of recovery of *Blastocystis* hominis in clinical practice	Zuckerman		1990	USA
7	13677378	Clinical characteristics and endoscopic findings associated with *Blastocystis* hominis in healthy adults	Chen	292	2003	Taiwan
8	10414382	Prevalence and clinical relevance of *Blastocystis* hominis in diverse patient cohort	Cirioni O.	1,216	1999	Italy
9	9158042	Epidemiologic survey of *Blastocystis* hominis infection in Japan	Horiki	6,476	1997	Japan
10	8545396	*Blastocystis* hominis: a common commensal in the colon. Study of prevalence indifferent populations of Paris	Junod	7,677	1999	France
11	16105126	No correlation between clinical symptoms and *Blastocystis* hominis inimmunocompetent individuals	Leder	2,800	2005	Australia
12	2401797	*Blastocystis* hominis: epidemiology and natural history	Senay		1990	Canada
13	7578767	Is *Blastocystis* hominis a cause of diarrhea in travelers? A prospective controlled study in Nepal	Shlim	301	1995	Nepal
14	2218447	Etiology of diarrheal diseases in immunocompetent and HIV-positive patients	Steinman	206	1990	Germany
15	10816147	Enteropathogens in adult patients with diarrhea and healthy control subjects: a 1-year prospective study in a Swedish clinic for infectious diseases	Svenungsson	1,053	2000	Sweden
16	8515120	*Blastocystis* hominis: prevalence in asymptomatic versus symptomatic hosts	Udkow		1993	USA

All studies identifying *Blastocystis* as pathogenic were performed on individuals from affluent countries. Most of the studies characteristics of *Blastocystis* infection which would seem to be unlikely in a pathogen. However, most of those characteristics had previously reported in *E. histolytica* and *G. intestinalis* patients

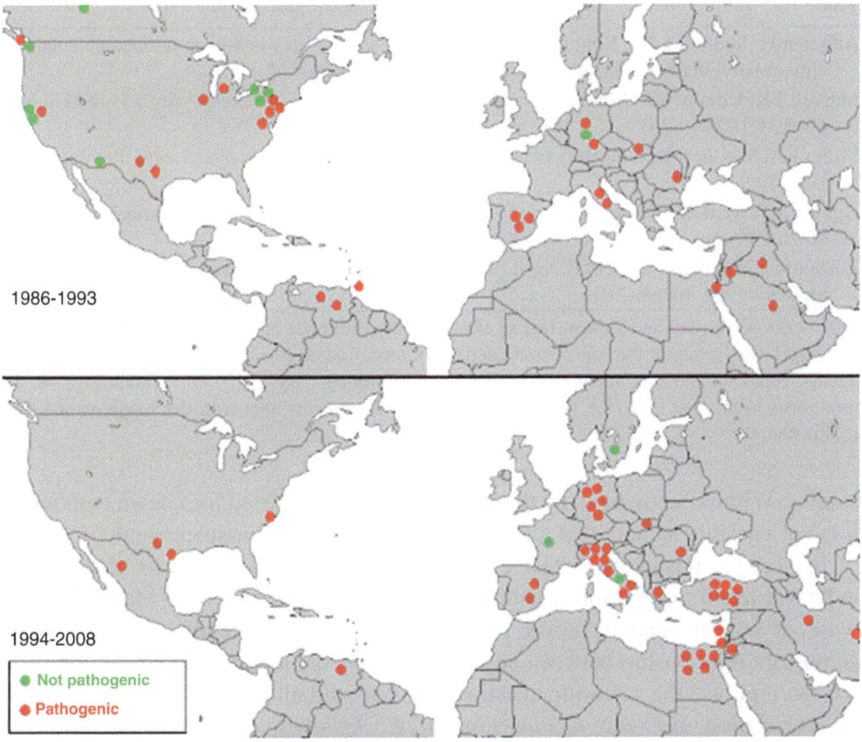

Fig. 7.5 Research conclusion in *Blastocystis* studies during two time periods. Most studies identifying *Blastocystis* as nonpathogenic were published between 1986 and 1993, and in those studies researchers generally identified characteristics of *Blastocystis* that were common to *G. intestinalis* and *E. histolytica*. After 1994, those studies became increasingly rare. Research came to a halt in the USA, possibly due to a policy change at the US National Institutes of Health, where staff members began informing potential *Blastocystis* researchers that no evidence existed to suggest *Blastocystis* was pathogenic

But the same property had been reported for *G. intestinalis* and *E. histolytica* in two papers studying patients in the same region a few years earlier:

> No correlation between symptoms and the presence of absence of infection <with Giardia lamblia and *Entamoeba histolytica*> could be detected
>
> –*Intestinal Parasitic Infections in Homosexual Men at a San Francisco Health Fair*, 1983
>
> No relation was seen between the presence or absence of gastrointestinal symptoms and infection with pathogenic protozoa.
>
> –*Intestinal protozoa in homosexual men of the San Francisco Bay area: prevalence and correlates of infection*, 1984

Following the rules of logic, if correlation with symptoms in humans is to be used as a litmus test for pathogenicity, one would need to understand why *G. intestinalis* and *E. histolytica* were uncorrelated with symptoms in those regional

Table 7.4 Studies cited by the US Center for Disease Control *Blastocystis* Factsheet, 1991–2010

Albrecht H, Stellbrink HJ, Koperski K, et al. *Blastocystis hominis* in human immunodeficiency virus-related diarrhea. Scand J Gastroenterol 1995;30:909–914
Markell EK, Udkow MP. *Blastocystis hominis*: pathogen or fellow traveler? Am J Trop Med Hyg 1986;35:1023–1026
Miller RA, Minshew BH. *Blastocystis hominis*: An organism in search of a disease. Rev Infect Dis 1988;10:930–938
Udkow MP, Markell EK. *Blastocystis hominis*: prevalence in asymptomatic versus symptomatic hosts. J Infect Dis 1993;168:242–244
Zuckerman MJ, Watts MT, Ho H., et al. *Blastocystis hominis* infection and intestinal injury. Am J Med Sci 1994;308:96–101

The works of Markell and Udkow had an extraordinary influence on US thought, to the extent that their work was virtually the sole source cited for information on *Blastocystis* until 2010, when that agency replaced the fact sheet and updated the references. By 2010, the average age of the references in this list was 18.6 years. Copies of both fact sheets appear in the supplemental files to this chapter

studies. Without knowing why the lack of correlation existed for known pathogens, it is not possible to state that a lack of correlation proves nonpathogenicity.

In this case, it is likely that Dr. Markell was aware of the two papers noted above, as he was listed as the primary author for both of them. In reviewing Markell's other work, it appears that he never published a paper that identified a statistically significant relationship between any enteric protozoal infection and symptoms. So we can rephrase the scientific question to, "Why did all of Markell's studies show no correlation between *G. intestinalis*, *E. histolytica*, and *Blastocystis* and symptoms?" Identification of cases where the same property is interpreted differently depending on researcher goals suggests that researcher interpretation has played a large role in study outcome.

There are, in fact, many counterintuitive findings that exist for enteric protozoal infections. In our review paper, we included a collection of these in an appendix called "Characteristics of Known Enteric Pathogens and Diseases." A copy of that appendix is included in the supplementary material for this chapter.

7.6 Analyzing Blastocystis Controversy

Our systematic analysis showed that most of the written controversy about *Blastocystis* occurred in the late 1980s and early 1990s, when a number of US physicians were publishing papers on behavior in *Blastocystis* thought to be incompatible with pathogenicity. The controversy peaked between 1988 and 1990, with researchers trading angry letters in the Journal of Clinical Microbiology. That exchange appeared to be the origin of the notion of *Blastocystis* as a controversial pathogen (Markell and Udkow 1990a,b; Rosenblatt 1990; Zierdt 1991).

Dr. Edward K. Markell, a physician with a health maintenance organization (HMO) in Oakland, California, and coauthor of the venerable Markell's *Medical Parasitology*, became particularly well known for his efforts in this area. Dr. Markell passed away in 1998, and *Blastocystis* appears to have followed him to the grave:

> He strongly believed that Blastocystis was not a pathogen, and he led virtually a one-man campaign to draw attention to its coincidental association with disease.
>
> In Memorian, Edward K. Markell, Parasitology Research, 1999 (Editor 1999)

Markell's work can be shown to be extraordinarily influential on *Blastocystis* opinion in the USA, to the extent that US federal policy from the mid-1990s to around 2010 was largely based on his research. The NIH's Pubmed database shows Markell authored five works on *Blastocystis*, but only two of those were research studies. The other three works were letters where Markell praised researchers who duplicated his findings, or attacked those who contradicted them. Despite this, both of those studies earned a place in the US Center of Disease Control's factsheet on *Blastocystis*, and one of those studies remain among the most cited *Blastocystis* papers of all time (Table 7.5).

Those five studies were the only ones cited on the US CDC Fact Sheet, which appeared on that organization's web site with few changes from February 2000 to March 2008, according to data archived from that organization's web site (CDC 2000). The agency added four additional references in late 2008, one of which questioned *Blastocystis* pathogenicity, two of which were performed in Zambia and China, and one which described treatment of children in Mexico with the drug nitazoxanide (CDC 2008). The agency did not include any study suggesting *Blastocystis* could cause illness in immunocompetent adults in developed Western countries until 2010, when an entirely revised fact sheet was released. That fact sheet excluded all five of the original studies cited in 2000. The new revision represented the first time when clinical studies of immunocompetent adults performed in the twenty-first century appeared in the fact sheet (CDC 2011). Copies of the original 2000 version and the 2010 revision are included in the supplementary material for this chapter.

In developing decision-making methodologies, there can be value in understanding if the methodology used in selecting research studies may be producing a bias in favor of a particular viewpoint. This can be done by calculating metrics for the research studies selected for an analysis, and determining if those studies are more likely to exhibit a property which could skew the conclusion. In the case of the five studies in Table 7.4, there are several metrics we can calculate, which might help identify a more neutral process for study selection. For example, in 2010 when the CDC data sheet was updated, the average of five studies identified by the sheet was over 18 years.

7.7 Assessing Confidence and Overconfidence in Medical Findings

Collecting statistics about researcher methods can help prevent cultural biases from influencing study conclusion, and later public policy. This process can also be performed in a way which is "conclusion-neutral" in any specific scientific

Table 7.5 The ten most frequently cited *Blastocystis* papers, based on a search for the word "*Blastocystis*" in the title of all journal articles indexed in the scientific collections of Google Scholar, performed on May 29, 2012

N	Number of times cited	Year of publication	First author	Title	Journal
1	288	1996	Dunn	Blastocystis hominis revisited.	Clinical Microbiology Reviews
2	222	1991	Zierdt	Blastocystis hominis—past and future	Clinical Microbiology Reviews
3	127	1986	Sheehan	Association of Blastocystis hominis with signs and symptoms of human disease	Journal of Clinical Microbiology
4	135	1993	Doyle	Epidemiology and pathogenicity of Blastocystis hominis	Journal of Clinical Microbiology
6	134	1993	Boreham	Blastocystis in humans and animals: morphology, biology, and epizootiology	Advances in Parasitology
7	111	1986	Markell	Blastocystis hominis: pathogen or fellow traveler?.	American Journal of Tropical Medicine and Hygiene
8	109	1967	Zierdt	Protozoan characteristics of Blastocystis hominis.	American Journal of Clinical Pathology
9	108	2002	Tan	Recent advances in Blastocystis hominis research: hot spots in terra incognita	International Journal for Parasitology
10	104	1997	Clark	Extensive genetic diversity in Blastocystis hominis	Molecular and Biochemical Parasitology

Google Scholar is a publicly accessible database of scientific literature which, in addition to Pubmed, is commonly cited in statistical studies of scientific literature. It represents one of the largest publicly available collection of citations to such literature. Google Scholar differs from the search engine Google in that the Google Scholar does not index web pages, and instead indexes scientific books and papers.

controversy. That is, we can agree to refrain from using certain practices that would be expected to produce an incorrect decision, and we can create those guidelines before the controversy begins. For example, we might agree that a list of references should not over-represent the work of one particular author. We might agree that we should not rely exclusively on very old studies, because that could also misrepresent the current level of understanding concerning a scientific topic. We might even want to favor studies that are more recent, since science is often viewed as an iterative process, and newer studies are likely to be constructed and reviewed by individuals who have more access to information about the subject.

In examining historical mistakes in medical science concerning microbial pathogenesis (Table 7.7), it can be seen that errors frequently originate from overconfidence in research findings from a single author or group, even before that finding has been reproduced by other researchers. Overconfidence in the lack of an association between a microbe and illness has been demonstrated as well (Table 7.8). Dr. Barry

Table 7.6 An account of Markell's life noted that he waged a "virtually one-man campaign against" the idea of *Blastocystis* being pathogenic, and his work was extraordinarily influential on medical thought in the USA (Table 7.4) This table shows items associated with Markell and Blastocystis from the NIH's Pubmed database.

Year	Type	Journal	Title	Details
September 1986	Study	Journal of Clinical Microbiology	*Blastocystis hominis: pathogen or fellow traveler?*	Asserted that another cause, such as IBS or another infection, could be found in all Blastocystis patients
March 1988	Letter	Journal of Clinical Microbiology	*Association of Blastocystis hominis with human disease*	Markell objects to publications by Sheehan, et al. and Pikula, et al. linking *Blastocystis* to illness in humans
May 1990	Letter	Journal of Clinical Microbiology	*Association of Blastocystis hominis with human disease?*	Markell objects to papers by Qadri, et. al. and Sheehan, et al. linking *Blastocystis* to illness in humans
June 1990	Letter	Western Journal of Medicine	*Blastocystis hominis*	Markell objects to publication by Babb, et al. identifying *Blastocystis* in joints of an immunocompromised patient
July 1993	Study	Journal of Clinical Microbiology	*Blastocystis hominis: prevalence in asymptomatic* versus *symptomatic hosts*	Markell identifies *Blastocystis* at a similar prevalence in symptomatic (12.8 %) and asymptomatic (11.5 %) patients, suggests that patients will be harmed by physicians who investigate *Blastocystis*
July 1995	Letter	Clinical Infectious Diseases	*Is There Any Reason to Continue Treating Blastocystis Infections?*	Markell praises study on expatriates living in Nepal that found *Blastocystis* infection at a similar prevalence in healthy and diarrheal patients

Markell only published two papers on *Blastocystis* before he died in 1998. Most (4/6) of Markell's *Blastocystis* citations are complaints he made against other researchers who published papers that conflicted with his findings, or praise for researchers who corroborated his work

Marshall, who shared the 2005 Nobel Prize for discovering the role of *H. pylori* in stomach ulcers, devoted a book to the subject of the number of times the relationship had been discovered before and lost to history due to researcher harassment by the medical community (Marshall 2002). The three-century delay between the discovery of *G. intestinalis* and its designation as a pathogen is examined in more detail in the next Chapter. A more recent example of the failure to recognize a pathogen would be

Table 7.7 Type 1 errors (incorrect attribution of a cause and effect) in medical studies

N	Journal	PI, Finding	Retracted/explained
Oct 2001	Science	Mikovits, XMRV found at significantly increased rate in CFS patients	Oct 2011 (Silverman et al. 2011)
Feb 1998	Lancet	Wakefield, MMR virus from vaccinations is found in colonic biopsies from autistic children, and causes autism	Mar 2004 Feb 2010 (full) (Dyer 2010)
Dec 1986	Infection & Immunity	Mirleman, *E. dispar* turns into *E. histolytica*	Dec 1993 (Clark and Diamond 1993)
1926	(Nobel Prize in Medicine)	Fibiger, Cancer can be induced in rats by infecting them with *Spiroptera neoplastica*	1930–1935 (Stolt et al. 2004)

In many cases, studies appeared to be revolutionary, and were embraced by portions of the medical community, but were subsequently found to result from experimental error. Applying simple criteria, such as requiring the finding to be independently produced by three different laboratories in three different countries, can help avoid false alarms

the decision in 2000 by the leadership of South Africa to promote the view that the HIV virus is only incidental in the development of AIDS.

A great deal of the controversy surrounding *Blastocystis* appears to have originated from the extraordinary level of confidence possessed by a few early researchers. In medical research, self-confidence may play a greater role in influencing opinion than it does in scientific research, where repeatability is more valued. As one interesting illustration of this phenomenon, Markell attacked a study where a researcher had identified *Blastocystis* cells in the joint fluid from a patient being treated with immunosuppressive drugs, insisting that the researcher had fabricated the data, or mis-identified the *Blastocystis* cells (Lee et al. 1990). But Markell had never performed a study on an immunocompromised patient. How could he have such confidence that the researcher was in error, since Dr. Markell's studies were exclusively performed on immunocompetent patients?

7.8 Decision Making Theory and Management Errors

Tables 7.7 and 7.8 suggest that there are a number of cases where a great deal of morbidity and mortality could have been avoided by following up on researcher findings in a timely manner. In reviewing this history, it may be tempting to assume that the decision makers of the time were corrupt or diabolical. But the difficulty in developing policies to address this problem lies in the fact that post-mortem investigations show little if any malfeasance on the part of the decision markers.

There is a body of research known as "Behavioral Decision Theory" which examines why this is true, especially as it relates to economic decisions and recent financial disasters (Bazerman and Moore 2009). This field of research is finding its way into medical science: the most frequently accessed paper in the Public Library of Science's journal PLOS-One is entitled, "Why Most Published Research

Table 7.8 Type 2 errors in medical decision making (failure to recognize a cause of a disease).

Organism	Duration of investigatory/indeterminate period (discovery in symptomatic humans to acceptance as pathogen)	Organization or individual credited with transitioning organism to pathogenic status
Vibrio cholerae	36 years (1849–1884)	Robert Koch
Giardia intestinalis	298 years (1681–1978)	US EPA Symposium on *Giardia* (Smith and Wolfe 1980)
Cryptosporidium sp.	10 years (1976–1997)	Various, including US Congress (Matukaitis 1997; Ortega et al. 1993)
Helicobacter pylori	91 years (1892–1984)	Marshall & Warren University of Western Australia (Marshall and Warren 1984)
Escherichia coli	36 years (1947–1983)	Riley US Center for Disease Control (Wells et al. 1983)
Norovirus	2 years (1971–1972)	NIH lab (Wyatt et al. 1974)
Blastocystis	100 years (1911–present)	Various

Errors of judgment are common in major infectious diseases, and disputes generate as many angry comments today asthey did in the 1850s, when experts opposed research identifying *Vibrio choleae* as the causative agent in cholera (Boorom 2009). The field of microbiology has not developed institutionally recognized methodologies for ascertaining pathogenicity. As a result, investigations into enteric pathogens can often span decades. In many cases, the issue of pathogenicity is not decided within the medical or scientific community, but rather by government legislation, as was the case for *G. intestinalis* and *Cryptosporidium* spp.

Findings Are False." An examination of this literature is of value, because it can tell us how we would need to approach decision making differently in medical science in order to avoid situations like those in Tables 7.7 and 7.8.

Harvard Law Professor Max Bazerman proposed the concept of "bounded ethicality" to explain this type of behavior in his book, "Judgment in Managerial Decision Making" (Bazerman and Moore 2009). As noted, in most egregious failures of management decision making, there is little evidence that the people making decisions were consciously harming others to promote their own interests. The committees of physicians that dismissed *Vibrio cholerae* as a cause of cholera in the 1840s had little to gain from the decision. Physicians who intimidated *H. pylori* researchers saw no gain from their behavior (Marshall 2002). Of the researchers listed in Table 7.7, Dr. Wakefield eventually lost his license to practice medicine in the UK, and a warrant was issued for the arrest of Dr. Mikovits, following allegations of the theft of research notebooks. As such, it is difficult to fit such researchers into a paradigm of individuals following well-designed plans to hoodwink the public.

Rather, the decision-making process told leaders that their actions were correct. Experiments on human subjects who are also experts in their field suggests that certain biases are so strong, that experts will consistently make the wrong decisions in certain situations, and that behavior can be repeatably demonstrated in human experiments. (Bazerman and Moore 2009). These biases are strongest in the following situations (Montier 2010):

1. The problem is ill structured and complex.
2. The information is incomplete, ambiguous, and changing.
3. Goals are ill-defined, shifting, or changing.
4. When stress is high due to time constraints or a high stakes.
5. When decisions rely on interaction with others.

In reviewing these factors, it is not difficult to see why physicians make mistakes about pathogenicity, as these factors exist in abundance in medical research. When we understand the idea of "bounded ethicality," the mistakes in Tables 7.7 and 7.8 can be seen in a different light. That is, researchers like Wakefield and Mikovits probably did not intentionally skew the findings of a study in order to produce a specific conclusion. In fact, the factors that lead them to publish and continually defend those findings may have operated at the sub-conscious level.

7.9 A Metric for Assessing Confidence in Research Findings

How can we develop systems that provide us with reliable and timely answers about microbial infections? One approach is to implement a structured decision-making process, as a substitute for unstructured decision making by experts. A researcher who calculates confidence intervals for relative risk from a contingency table uses a structured process, while a physician who draws a conclusion from a "gut feeling" is using an unstructured process. When a structured process replaces expert guidance, there is sometimes objection that a formula cannot replace the experience and wisdom of an expert. However, in many cases, an evidence-based structured process will out-perform unstructured decision making by experts, because such unstructured decision making is prone to bias (Montier 2010).

In the next chapters, we will discuss specific findings about *Blastocystis*, and we will use a structured decision making system to assess the level of confidence we should have in those findings. The following criteria are used to classify findings:

1. *High confidence findings* are those corroborated by at least three different research groups operating in three different countries, and where no contradictory studies exist, or where contradictory studies number $<10\ \%$ of studies with the positive finding.
2. *Medium confidence findings* are those reported in at least one study, the conclusions of which are not contradicted by another study. Additionally, the findings must either identify a phenomenon which is known to be true for all similar infectious diseases or which is supported by the presence of circumstantial evidence from several other studies.
3. *Findings which do not meet any of these criteria* occasionally work their way into the medical community, and these are discussed as well.

The following provides two examples of the application of these criteria:

1. ***"Is Blastocystis transmitted by contaminated water?"*** In 2000, researchers from the USA and Jordan had independently suggested an association between contaminated or untreated water and *Blastocystis* infection (Nimri and Batchoun 1994; O'Gorman et al. 1993). With just two studies, this could not be a High Confidence Finding. Since all highly prevalent pathogenic protozoa (*Cryptosporidium* spp., *G. intestinalis*, *E. histolytica*) are known to be transmissible via contaminated water, this would be classified as a "Medium Confidence Finding." An additional study appeared in 2004 from Thailand implicating waterborne transmission of *Blastocystis* infection (Leelayoova et al. 2004). Between 2007 and 2012 close to a dozen studies appeared in the literature from multiple countries, including ones which genotyped *Blastocystis* in humans and water samples (Leelayoova et al. 2008). As such, at the current time, transmission of *Blastocystis* by contaminated water would be considered to be a "High Confidence" finding.
2. ***"Can symptomatic and asymptomatic Blastocystis infection in humans be explained primarily by variation in pathogenicity of different Blastocystis subtypes?":*** As a second example, in 2006, some findings could be extended to suggest that symptomatic and asymptomatic infection may be dependent on different subtypes of *Blastocystis* (Tan and Suresh 2006a,b). However, the evidence for this was based on findings from one laboratory, which does not mean it was incorrect, but it may have been related to some other variable or process. Additionally, while parasite genotype has been important in differentiating *E. histolytica/dispar* infection, it has not been as significant in explaining symptomatic and asymptomatic in *G. intestinalis* or *Cryptosporidium* spp., so it does not meet the universality criteria described previously. Additionally, there was not a great deal of circumstantial evidence, for example, symptomatic and asymptomatic *Blastocystis* infection did not cluster in certain areas or families. Subsequent studies from multiple sites showed that the same subtypes of *Blastocystis* are present in symptomatic and asymptomatic patients (Dogruman-Al et al. 2009; El-Shazly et al. 2005), suggesting other factors may be more important in determining asymptomatic or symptomatic infection. As such, the proposal that symptomatic and asymptomatic infection is determined by *Blastocystis* subtype does not meet the criteria for either a high-confidence or a medium-confidence finding at this time.

The *Blastocystis* Research Foundation (BRF) began categorizing studies on *Blastocystis* infection in 2006 for our metastudy paper. The initial effort was limited to categorization of studies by researcher finding (pathogenic or nonpathogenic). We continued distributing the list of those two categories of studies, but we found that researchers were also interested in lists of other studies into specific areas of *Blastocystis*. In, we extended the original two-category list to include additional categories, with the complete list as follows (as of March 17, 2012):

Studies Identifying *Blastocystis* as Pathogenic or Correlated with Symptoms (157 studies)
Studies Identifying *Blastocystis* as Nonpathogenic or Uncorrelated with Symptoms (17 studies)
Studies Identifying *Giardia* and *E. histolytica* as Nonpathogenic, Uncorrelated with Symptoms, or Not Requiring Treatment (37 studies)

Studies of *Blastocystis* and Irritable Bowel Syndrome (IBS) (22 studies)
Studies Linking *Blastocystis* to Inflammatory Bowel Disease (12 studies)
Studies Describing the Physiological Effect of *Blastocystis* in Humans and Pathogenesis of Blastocystosis (16 studies)
Studies Describing How to Detect *Blastocystis* in Humans (31 studies)
Animal Models for *Blastocystis* Infection (19 studies)
Antibody Response to *Blastocystis* (19 studies)
Culturing *Blastocystis* (63 studies)
PCR Detection of *Blastocystis* (50 studies)
Blastocystis Treatment Studies and Reviews (46 studies)
Studies Identifying the Proportion of *Blastocystis* Monoinfections that are Symptomatic (3 studies)
Selected Studies on the Role of IL-8 and Nitric Oxide in *Blastocystis* infection, IBS, and *Helicobacter pylori* infection (23 studies)

7.10 High Confidence Findings

In population studies of groups of 10 or more adults from non-Asian countries, between 68 % and 100 % of all individuals with Blastocystis mono-infections report gastrointestinal symptoms: A study published in 2002 by a US researcher surveyed 2,896 stool samples collected in 2000 and found that 69 % (400/581) of the individuals the *Blastocystis* mono-infections reported symptoms (Amin 2002). A 2010 study noted that 63 % (12/19) of zookeepers studies were found to have *Blastocystis* infection and only *Blastocystis* infection (Parkar et al. 2010). All of those cases (100 %, 12/12) were symptomatic. And a study of 108 food handlers in Egypt who were mono-infected with *Blastocystis* found that 68.5 % were symptomatic (Fathy 2011).

Blastocystis can be transmitted by contaminated water: Researchers have identified an increased risk of *Blastocystis* infection in groups consuming untreated water in studies from the USA (O'Gorman et al. 1993), Thailand (Leelayoova et al. 2004), and China (Li et al. 2007b). Genotyping studies have shown the type of *Blastocystis* present in water matches the type in individuals infected with *Blastocystis* (Leelayoova et al. 2004; Li et al. 2007b). Viable *Blastocystis* has been demonstrated in sewage samples (Suresh et al. 2005) and in recreational water samples (Ithoi et al. 2011).

Blastocystis infection is found at an elevated rate in IBS patients from all non-Asian countries: The first study that suggested *Blastocystis* was causing IBS was published in 1997 by a University research group in Karachi, Pakistan (Hussain et al. 1997).[3] Six studies from five countries are summarized in Table 7.9. The relationship between *Blastocystis* and IBS does not appear in Thai patients (Surangsrirat et al. 2010; Tungtrongchitr et al. 2004).

[3] This finding was in contrast to Markell's 1986 paper, which indicated that *Blastocystis* patients could be diagnosed with IBS (Markell and Udkow 1986).

Table 7.9 *Blastocystis* is significantly correlated with irritable bowel syndrome (IBS) in all studies from non-Asian countries

No.	Year	Country	*Blastocystis* infection rate in IBS patients	*Blastocystis* in asymptomatic controls or general population
1	2011 (Jimenez-Gonzalez et al. 2011)	Mexico	31.1 % (14/45)	13.3 (6/45)
2	2010 (Yakoob et al. 2010a)	Pakistan	53 % (90/171)	16 % (25/159)
3	2010 (Dogruman-Al et al. 2010)	Turkey	76 % (13/21)	11.6 (5/43)
4	2005 (Windsor 2007)	UK	38 % ($n > 800$)	7 %
5	2004 (Yakoob et al. 2004b)	Pakistan	46 % (44/95)	7 % (4/55)
6	1999 (Giacometti et al. 1999)	Italy	18.5 % (15/81)	7.5 % (23/307)

This relationship does not appear to be present in Thailand, based on data from two studies (Surangsrirat et al. 2010; Tungtrongchitr et al. 2004). Variations in prevalence of infection in IBS patients between studies may be largely due to different detection methods used, which included PCR detection, stool culture, and less-sensitive staining techniques

Experimental infection of laboratory animals with Blastocystis will produce gastrointestinal or physiological symptoms, but only if large numbers of cysts are used: One of the first reports of symptoms in a laboratory animal following experimental infection was published by a research group from Singapore, describing experimental infection of mice in 1997 (Moe et al. 1997). This was followed by additional studies in mice from China in 2005 and 2006 which reported severe symptoms and death (Yao et al. 2005; Zhang et al. 2006). A study followed on experimental infection of rats with 40 million cysts in Egypt in 2008 which reported severe inflammation and precancerous polyps (Hussein et al. 2008). A 2010 study on murine infection noted no changes mice inoculated with 100 or 100,000 cysts, but did note significant changes in mice inoculated with 47 million cysts (Elwakil and Hewedi 2010) Studies describing high levels of oxidative stress in urine samples from experimentally infected rats were reported by a Malaysian group in 2010 (Chandramathi et al. 2010b), and the group noted that urine samples from humans exhibited a similar phenomenon in humans (Chandramathi et al. 2009). A 2009 study from a Japanese group noted changes in cytokine response in experimentally infected rats, but no histological changes in biopsies of rats. The animals were inoculated with 100,000 cysts (Iguchi et al. 2009).

The prevalence of Blastocystis infection varies substantially between different geographies and different groups within a single country: The largest study of *Blastocystis* infection in different groups within the same country was performed in China and published in 2010 (Li et al. 2007a) and noted a prevalence varying from a low of 1.9 % in the Shanghai municipality to a high of 32.6 % in Yunnan province.

This is the only large population study ($n = 2321$) performed using a detection technique (stool culture) considered by most researchers to be reliable. Studies from additional countries showing substantially different prevalences of infection are shown in Table 7.9. A study on *Blastocystis* in zookeepers from Australia and Europe reported a prevalence of 63 % (12/19) in the zookeepers (Parkar et al. 2010), which was higher than the prevalence in the zoo animals (!) All individuals infected with *Blastocystis* in that study reported gastrointestinal symptoms.

It should be noted that researchers sometimes misrepresent the frequency of detection of *Blastocystis* in clinical samples as the prevalence in the population. However, this practice is incorrect, as clinical samples usually come from patients with an illness, and such samples will often contain *Blastocystis* much more often than samples from the general population. Additionally, as clinical detection methods have a sensitivity of <40 % (Dogruman-Al et al. 2010; Leelayoova et al. 2002; Stensvold et al. 2007), it is not reasonable to represent figures from those methods as a prevalence.

The diagnostic techniques commonly used in clinical settings fail to detect most Blastocystis infections: One of the first studies to note this relationship was published by a Thai research group in 2002, noting that the sensitivity of smear and concentration techniques was about 32.3 % compared to in vitro culture (Leelayoova et al. 2002). This finding was repeated in 2004 with a study on UK patients (Suresh and Smith 2004). Additional studies performed in Denmark noted that staining and concentration techniques in common use in Europe would detect about a third of the infections identifiable by polymerase chain reaction (PCR) or stool culture (Stensvold et al. 2006, 2007). A pair of US studies noted the identification of *Blastocystis* infection in a majority of patients with chronic gastrointestinal illness, even though most patients had been found to parasite-negative by clinical laboratories (Jones et al. 2009; Whipps et al. 2010). A study on Turkish patients published in 2010 noted that trichrome staining would detect only about a third of the infections detectable by PCR and stool culture (Dogruman-Al et al. 2010). A 2010 comparison study in which the same stool samples were analyzed by multiple European reference laboratories found that agreement between laboratories for *Blastocystis* detection was the worst for any intestinal parasite examined (Utzinger et al. 2010).

In Blastocystis mono-infections, the following symptoms are among the most common reported by Blastocystis patients: diarrhea, abdominal pain, vomiting, flatulence, and fatigue: This symptomatic distribution was first reported in a study of 239 mono-infected patients from Saudi Arabia in 1989 (Qadri et al. 1989). A similar distribution was found in a study of *Blastocystis* mono-infection in 39 children in the USA in 1993 (O'Gorman et al. 1993), as well as a study of 38 mono-infected schoolchildren in Jordan in 1993 (Nimri 1993). Similar symptoms were noted in a study of 23 mono-infections from Egypt in 2005 (El-Shazly et al. 2005), and in 108 mono-infections reported from Libya in 2007 (Al-Fellani et al. 2007), and a detailed molecular study of *Blastocystis* subtypes, coinfections, and symptoms in 92 patients from Denmark in 2009 (Stensvold et al. 2009).

Blastocystis subtype 3 is the most common type of Blastocystis infection in most Western countries: The first subtyping of isolates from a Western country was performed in Denmark in 2006 and identified subtype 3 as the most prevalent variant (Stensvold et al. 2006). This was followed by similar findings from a Greek study in 2008 (Menounos et al. 2008), a US study in 2009 (Jones et al. 2009), a French study in 2009 (Souppart et al. 2009), a Mexican study in 2011 (Jimenez-Gonzalez et al. 2011), and a Swedish study in 2012 (Forsell et al. 2012). Subtype 3 is also the most common *Blastocystis* type found in Egypt (Hussein et al. 2008; Souppart et al. 2010) and China (Li et al. 2007a). Different subtype distributions have been found in a 2012 Australian study (Nagel et al. 2012), where Subtype 1 was the most common (45 %) and subtype 3 was as common as subtype 4 (36 %). In Nepal, 86.4 % of *Blastocystis* infections were found to be made up of subtype 4. Additionally, in Denmark, subtype 4 was found to be the most common subtype among patients with acute (rather than chronic) diarrhea (Stensvold et al. 2011).

In Western countries, age is a risk factor for the development of long-term illness with Blastocystis infection. That population consists primarily of individuals over 30: The average age of individuals with long-term illness associated with *Blastocystis* mono-infection is usually significantly greater than the average age of the population studied. This relationship holds true in studies conducted in Turkey (Dogruman-Al et al. 2010), Mexico (Jimenez-Gonzalez et al. 2011) and Pakistan (Yakoob et al. 2010b), with a typical average age of 41 years reported in such studies.

Blastocystis upregulates production of the inflammatory cytokine IL-8: This was first reported in 2001 in a German study of *Blastocystis* cells co-cultured with human colonic cells (Long et al. 2001), and then again in two studies in 2008 from Singapore labs which noted differences between subtypes in IL-8 upregulation behavior (Mirza and Tan 2009; Puthia et al. 2008). Similar findings were reported in 2010 from a co-culture study performed in Malaysia (Chandramathi et al. 2010a). An interesting finding noted in the next section is that individuals with a polymorphism which upregulates production of the cytokine IL-8 are much more likely to show symptoms when infected with *Blastocystis* (Olivio-Diaz, 2012).

7.11 Medium Confidence Findings

In adult Blastocystis patients, the duration of infection approaches or exceeds 1 year and may last indefinitely in some individuals: A Japanese study was the first to report multiyear *Blastocystis* infection, which was accompanied by a steadily increasing immune response to the infection (Kaneda et al. 2000). Studies from the USA and other countries have related *Blastocystis* infection to long-term illness in patients, which has been found to be of a duration of 5 years or more (Jones et al. 2009). A Thai study followed adult caregivers at an orphanage and found that children cleared *Blastocystis* in an average of 2.6 months (median age 10 months), while adults cleared *Blastocystis* in an average of 10 months (median age 37 years) (Pipatsatitpong et al. 2012).

Expression of symptoms in Blastocystis infection may depend significantly on the host's genetic characteristics: The first study to note a specific host genetic effect in *Blastocystis* infection was performed on 45 patients and 45 controls in Mexico and found that two mutations that up-regulate IL-8 and IL-10 production were significantly associated with symptomatic expression in hosts, with the IL-8 characteristic potentially accounting for 45 % of the etiological fraction of illness (Olivo-Diaz et al. 2012). Relationships between host genetic traits and symptomatic expression in microbial illness can be identified for most major pathogens. The most common example may be that heterozygous carriers of the sickle cell trait exhibit resistance to extreme symptoms in malaria. Blood type A has been found to confer protection against infection with *E. histolytica* (Haque et al. 2003). Hosts with blood types A and AB may be more likely to experience symptomatic infection with enterotoxic *E. coli* (ETEC) (Qadri et al. 2007).

The antimicrobial therapies commonly used to treat Blastocystis infection may fail to eradicate the infection in most or all cases, especially in Western countries: The development of resistance to metronidazole was first reported in the mid-1990s (Zaman and Zaki 1996), and isolates from "IBS" patients were tested for antimicrobial resistance for the first time in a study from Pakistan in 2004 (Yakoob et al. 2004a). A poster presented in 2007 at the 15th United European Gastroenterology Week in Paris by Borody and Wettstein noted that most Australian *Blastocystis* patients appeared to be receiving no benefit from metronidazole treatment, and described a combination of drugs which provided better outcomes. A 2010 review paper, aptly named, "Antimicrobial therapy in *Blastocystis*: Reality or Delusion," critically examined the evidence surrounding the use of antimicrobials in treating *Blastocystis* patients (Stensvold et al. 2010). Surprisingly, there are very few studies in which patients who are treated for *Blastocystis* infection are subsequently tested for the infection to determine if the therapy can be said to eradicate the microbe. One of the first such studies was performed on 11 Australian patients in 2011, and it reported that every (11/11) patient with long-term *Blastocystis* symptoms who was treated with 400 mg of metronidazole for 11 days was still positive for *Blastocystis* infection at the end of treatment, with *Blastocystis* detected by PCR (Nagel et al. 2012).

7.12 Findings Which Do Not Meet any of These Criteria

Blastocystis only produces symptoms in immunocompromised patients: It is not clear where this concept originates. Overall, when we reviewed *Blastocystis* literature in 2008, we identified 86 studies pointing to symptomatic *Blastocystis* infection in immunocompetent patients, but only about half that number in immunocompromised patients. That trend has continued post-2008, with about 80 % of the current studies addressing immunocompetent patients, or the general population, and only a few studies focusing on HIV, cancer, and organ transplant patients. A 1990 study of

130 Canadian *Blastocystis* patients noted that only a few were immunocompromised (Doyle et al. 1990).

The origin of this idea may be related to the fact that some other microbes only cause symptoms in HIV and patients undergoing chemotherapy. The physician thus extrapolates that finding to assume that variability in symptomatic expression for all microbial infections is due to HIV infection and cancer treatment. Alternately, it may be the case that HIV and cancer patients visit physicians more often, and have more tests performed, and the discovery of *Blastocystis* in those patients establishes a mental correlation between *Blastocystis* infection and host immune status.

Blastocystis infection is easily treatable in all patients with metronidazole: In the few studies that have followed up with patients treated with metronidazole using sensitive detection techniques, that drug has been shown to have a high failure rate, particularly in patients in developed countries (Nagel et al. 2012; Stein 2007).

7.13 Is Giardia Good For You? Understanding Population Studies with Analytical Methods

A 2006 paper on *Blastocystis* from the London School of Hygiene and Tropical Medicine (Scicluna et al. 2006) noted that:

> Most population-based studies find no difference between rates of infection in symptomatic and asymptomatic individuals. In contrast, in individual infections a strong case can be made for the organism being the cause of disease.

Why is this? And is the phenomenon unique to *Blastocystis*? A short time spent reviewing papers on established pathogenic gastrointestinal protozoa will show that many researchers have performed population studies and found that these organisms can, in some populations, show no correlation or little correlation with illness (Janoff et al. 1990; Markell et al. 1983, 1984). Patients with *G. intestinalis* infection have even been reported to be healthier than those without such infection (Veenemans et al. 2011). Does this mean that protozoal infections are good for you?

Surprisingly, there are few papers in which a researcher has begun with known properties of existing pathogens, and determined if it should be possible to consistently demonstrate pathogenicity in all epidemiological situations. That is, can the known properties of established gastrointestinal protozoa account for researcher findings that show no correlation between *G. intestinalis* infection and symptoms?

Consider the Relative Risk and statistical significance (Fisher's Exact Test or Chi square) of a 2 $\times$ 2 contingency table representing a population of 1,000, all of whom are infected with *Giardia*, but only 50 % of whom show symptoms. We can assume that the asymptomatic individuals have acquired or innate immunity, for example.

	Giardia+	*Giardia−*	Total
Diarrhea	500	0	500
No diarrhea	500	0	500
Total	1,000	0	1,000

We can now show that in this population, the relative risk of having diarrhea is the same, whether the patient is infected with *G. intestinalis* or not. So since the RR $= 1$, the contingency table shows no relationship between *G. intestinalis* and diarrhea.

However, lack of correlation does not prove lack of causation, and in this case, we can see that if symptomatic expression is dependent on a host genetic or immunological factor, correlation studies may not demonstrate pathogenicity for a particular microbe. As a further examination, consider the effect of adding 250 individuals, all of whom are infected with *Cryptosporidium* sp., and all of whom have diarrhea.

	Giardia+	*Giardia−*	Total
Diarrhea	500	250	750
No diarrhea	500	0	500
Total	1,000	250	1,250

We now see that 67 % (500/750) of the patients with diarrhea test positive for *G. intestinalis*, but 100 % of the patients who are healthy have *G. intestinalis*. Thus, people with *G. intestinalis* are healthier than those without *G. intestinalis*! The example can be repeated with larger numbers of patients to create a "landmark" study. Such studies which might involve thousands of patients are just as susceptible to confounding variables as smaller studies.

The results do not tell us anything about whether we should treat patients, or whether *G. intestinalis* is causing illness. If we refuse to treat *G. intestinalis* patients based on this study, we leave up to half of the population with a treatable type of diarrhea. It is worth noting that once the correlation calculation is performed, many researchers will stop analyzing the results, and conclude that *G. intestinalis* (or *Blastocystis*) is either harmless or even healthful:

> Healthy day care children with asymptomatic *Giardia* infection show no disadvantage and perhaps even an advantage in nutritional status and freedom from other illnesses. (Ish-Horowicz et al. 1989)
>
> An additional study examining Tanzanian children noted that children who tested positive for *Giardia intestinalis* were less likely to have diarrhea, but that relationship was lost when the children were treated with a nutritional supplement. (Veenemans et al. 2011)

To focus on the question of pathogenicity, it is necessary to remove confounding variables:

1. The inclusion patients with other infectious diseases that produce the same symptoms as the microbe under study will weaken statistical correlation, or even reverse it.

2. If adaptive or innate immunity can exist for the microbe under study, including patients who are immune to the infection will weaken the correlation. Limiting the study group to patients who are older or have specific genetic traits can improve correlation figures for *Blastocystis* (Olivo-Diaz et al. 2012).

7.14 Acquired Immunity and Peak Shifting

Researchers have noted seemingly enigmatic behavior in both *E. histolytica* and *G. intestinalis*, as they discover groups with a high infection rate, and few symptoms. This effect has largely been ignored in public health efforts, which generally focus on reducing exposure, rather than building immunity through vaccination. It is necessary to understand this effect in order to properly assess the significance of potential new enteric microbes and also to understand potential effects of reducing exposure to these microbes. One remarkable study followed 82 schoolchildren for months and showed they exhibited no symptoms from *G. intestinalis* infection:

> We prospectively evaluated excretion of *Giardia lamblia* in children in day care centers in Houston by conducting two prevalence studies of 600 children enrolled in 30 day care centers, and an 18-month longitudinal study in 82 children in one center. In the two prevalence surveys, *Giardia* cysts were identified in 72 (21 %) and 67 (26 %) children, respectively, who provided stool specimens. Trophozoites were found in 15 (4 %) and 8 (3 %), respectively. There was no correlation between the frequency of recent diarrheal episodes and the finding of *Giardia*. In the longitudinal study, cysts were detected in stool specimens from 27 (33 %) of the 82 children at least once during the survey. Twelve children had *Giardia* cysts in weekly stool specimens for a mean of 6.2 ± 1.2 months and trophozoites for 3.3 ± 1.2 months. The number of enteric symptoms observed in children and the classification of nutritional status based on monthly height and weekly weight measurements did not differ significantly when infected and noninfected children were compared.
>
> –Occurrence of *Giardia lamblia* in children in day care centers, 1984 (Pickering et al. 1984)

> *Giardia lamblia* infection was identified in 33 of 89 (37 %) 3-month-old to 3-year-old children who were followed with monthly stool examinations for up to 12 months in a day care center. The infection was mainly asymptomatic and usually associated with prolonged carriage of the parasite. There were no significant differences for height and weight achievements and mean hemoglobin values between *Giardia*-positive and *Giardia*-negative children. However, *Giardia*-positive children tended to achieve higher weight and height for age than *Giardia*-negative children; weight for age was above the 50th percentile in 69 % of *Giardia*-positive vs. 40 % of *Giardia*-negative children (alpha = 0.01). *Giardia*-positive children tended to have fewer symptoms related to the gastrointestinal and respiratory tracts as recorded by a weekly questionnaire. Lactase deficiency was detected by breath hydrogen testing in 8 of 26 *Giardia*-positive vs. only 1 of 21 *Giardia*-negative children (*P* less than 0.02). Healthy day care children with asymptomatic *Giardia* infection show no disadvantage and perhaps even an advantage in nutritional status and freedom from other illnesses.
>
> –Asymptomatic giardiasis in children, 1989 (Ish-Horowicz et al. 1989)

> We conducted a point prevalence survey for enteric protozoa in 205 institutionalized orphans 1-61 months of age in Bangkok, Thailand. *Cryptosporidium* was identified in 17 children (8%), *Giardia lamblia* in 42 (20%), and 3 children (1%) had both parasites. At the

time of diagnosis, diarrheal symptoms were present in a minority of subjects: 36% of children with *Cryptosporidium* alone, 10% with G. lamblia alone, and in 20% of those with neither parasite...Although neither infection with *Cryptosporidium* nor G. lamblia was consistently associated with acute diarrheal symptoms, *Cryptosporidium* was more often associated with depressed acute nutritional status than *G. lamblia*.

–Endemic *Cryptosporidium* and *Giardia lamblia* infections in a Thai orphanage (Janoff et al. 1990)

The most interesting example of asymptomatic *E. histolytica* infection may come from a 1925 paper from Harvard Medical School professor Dr. SF Chiang, who infected rats with *E. histolytica* from his own stool samples:

Material for the second series of experimental infections was furnished by Dr. S., he himself being a healthy carrier of *E. histolytica*, *E. coli* and *Endolimax nana*.

–The Rat as a possible carrier for *E. histolytica* (Chiang 1925)

The rats did, in fact, develop illness. Can these studies help us understand some of the mystery behind why some people show symptoms in *Blastocystis* and others do not? Why does the long-term morbidity keep showing up in Europe but not in Africa or Asia (so far)?

One commonality of all these studies is that they focus on individuals in hyper-endemic environments, where exposure is occurring. We can see a similar phenomenon in a modern study of *E. histolytica* in Bangladeshi children:

The contribution of amebiasis to the burden of diarrheal disease in children and the degree to which immunity is acquired from natural infection were assessed in a 4-year prospective observational study of 289 preschool children in an urban slum in Dhaka, Bangladesh. *Entamoeba histolytica* infection was detected at least once in 80%, and repeat infection in 53%, of the children who completed 4 years of observation. Annually there were 0.09 episodes/child of *E. histolytica*-associated diarrhea and 0.03 episodes/child of *E. histolytica*-associated dysentery.

–*Entamoeba histolytica* infection in children and protection from subsequent amebiasis (Haque et al. 2006)

In some pathogens, groups with the greatest exposure also experience the greatest morbidity. This is clearly true for exposure to *Mycobacterium tuberculosis* and the HIV virus. But a pattern can be seen for gastrointestinal protozoa in which individuals who are exposed frequently or continuously do now show severe manifestations of the illness. This is one reason researchers have focused on the development of a vaccination for *E. histolytica* (Haque 2006).

Studies showing that enteric protozoa as uncorrelated with symptoms are often conducted on populations from day cares, orphanages, prisons, or on gay men, where exposure is common (Table 7.10). On the other hand, studies of individuals with infrequent exposure show illness, which is often severe. Cases like this include vacationing travelers from developed countries, or community exposures in areas that usually have uncontaminated water supplies (de Lalla et al. 1992).

Individuals in a Thai community are found to maintain high levels of immunity to *G. intestinalis* throughout their lifetime, while individuals in the USA were found to exhibit declining levels of immunity after age 19 (Fig. 7.6). One interpretation of

Table 7.10 Many studies suggest that patients with *G. intestinalis* or *E. histolytica* are just as healthy as those without such infections

Year	Study group	Region	Organism	Finding
1954	Prisoners	Southern USA	*G. intestinalis*	Experimental infection of prisoners with *G. intestinalis* produces no clinical manifestations of giardiasis (Rendtorff 1954)
1954	Prisoners	Southern USA	*E. histolytica*	Experimental infection of prisoners with *E. histolytica* produces no clinical symptoms of amoebiasis (Rendtorff and Holt 1954)
1986	Gay Men	UK	*E. histolytica*	*E. histolytica* appears to be a "commensal" in gay men (Allason-Jones et al. 1986)
1984	Children in daycare	Houston	*G. intestinalis*	No correlation between symptoms and infection (Pickering et al. 1984)
1989	Children in day care	Israel	*G. intestinalis*	Children with *G. intestinalis* appear healthier (Ish-Horowicz et al. 1989)
1990	Children in Thai orphanage	Thailand	*G. intestinalis* *Cryptosporidium* spp.	No correlation between diarrhea and *G. intestinalis* (Janoff et al. 1990)
2011	Children	Tanzania	*G. intestinalis*	*G. intestinalis* provides protection against diarrhea unless a food supplement used (Veenemans et al. 2011)

These studies are often performed on population where the infection is endemic, suggesting the possibility that acquired immunity may play a significant role in asymptomatic infection

the data would suggest that public health efforts do not prevent all exposures to gastrointestinal protozoa, but rather have the effect of delaying the first exposure and creating more time between exposures. This produces an epidemiological characteristic known as peak shifting, where the peak in prevalence vs. age occurs at a later time. For some microbial infections which are more severe when first exposure occurs later in life, the improved efforts at sanitation may have the paradoxical effect of increasing morbidity in the population by reducing the frequency of exposure, but increasing the severity of illness following an exposure.

7.15 How Much is *Blastocystis* Costing Communities?

A number of papers have provided cost estimates for outbreaks of acute illness due to food-borne bacterial and *Cryptosporidium* spp. (Corso et al. 2003, Batz et al. 2012). Comparable studies for *Blastocystis* would be complicated by the clinical use of low-sensitivity diagnostic tests, and a lack of reporting mechanisms for infections.

A more feasible approach would be to use existing cost estimates for IBS and adjust those by the percentage of cases which are due to *Blastocystis* infection. The latter factor might be calculated by identifying the percentage of IBS patients who

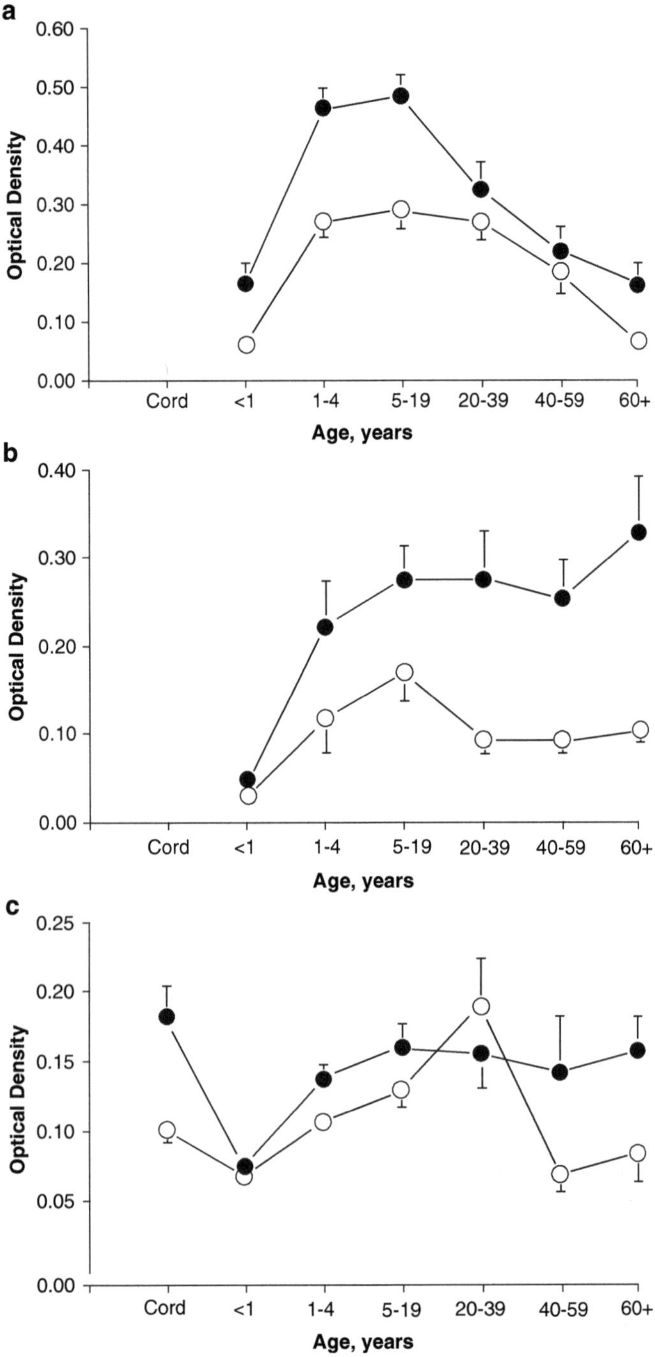

Fig. 7.6 Average serum antibody to *G. intestinalis* antigens response for 210 residents of Denver, Colorado versus 207 residents of Soongern, Thailand. In the USA, antibody response declines substantially after age 19, which is not seen in Thailand. If illness is more severe or long lasting in immunologically naïve adults over 30, a counterintuitive trend may develop, where the population with a lower prevalence of an infection sees more incidences of severe illness

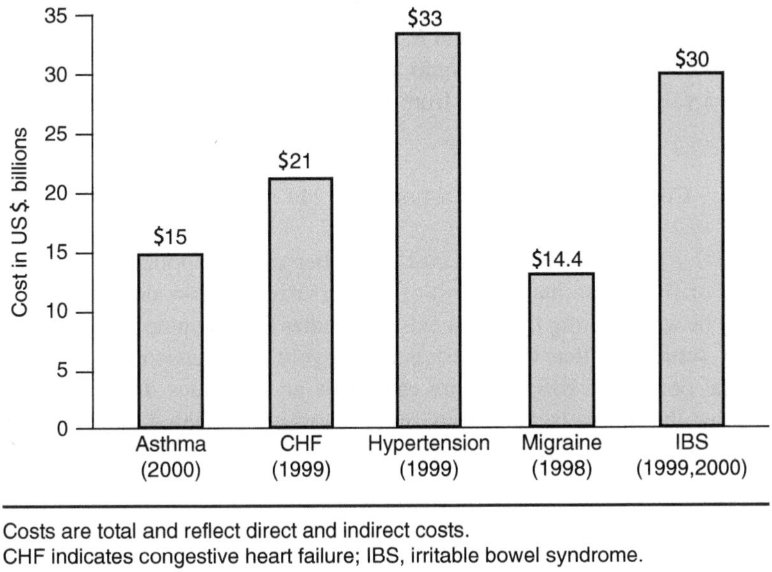

Fig. 7.7 Comparison of the total costs for various long-term illnesses in the USA (Cash et al. 2005). Existing studies suggest that 24.4–46 % of IBS morbidity may be attributable to *Blastocystis*, which would place the cost of *Blastocystis* infection at between $7.3 and $13.8 billion in the USA

are infected with *Blastocystis* by the percentage of those infections which are expected to be symptomatic:

$$\text{Cost of Blastocystis} = P_{\text{IBS}} \times P_{\text{symp}} \times \text{cost of IBS}$$

Where P_{IBS} is the percentage of IBS patients who are infected with *Blastocystis* and P_{symp} is the percentage of *Blastocystis* infections that are symptomatic.

This formula accounts for the argument that some patients who are infected with *Blastocystis* do not show symptoms, so a portion of the *Blastocystis*-positive IBS patients could represent asymptomatic *Blastocystis* carriers who have another cause for the illness.

When studies using stool culture or PCR detection are considered, P_{IBS} has been reported as 31, 53, 76, 38, and 46 % in existing studies (Table 7.9), giving 48.8 % as an average.

Numbers for P_{symp} for non-Asian countries have been reported to be in the range of 69–100 % (see High confidence findings). One of those studies used a small sample, and two of those studies could have arguably been influenced by patient self-selection (Amin 2002; Yoshikawa et al. 2004), so an argument could be made that P_{symp} could be lower. However, in studies of other enteric protozoa in healthcare-seeking individuals, resolution of symptoms occurs in over 95 % of healthcare-seeking patients following eradication of the infection, suggesting that P_{symp} for healthcare-seeking patients may be higher than P_{symp} for the general population.

For example, P_{symp} for *E. histolytica* is estimated to be only 10 %, but in treatment study of patients with symptoms and *E. histolytica* infection, 80 to 90 % of patients report resolution of symptoms (Latonio, 1988).

Using a value of P_{symp} ranging from 50 to 95 %, combining the two coefficient gives:

$$\text{Cost of } Blastocystis \text{ infection} = (0.244 \textit{ to } 0.46)\, cost\ of\ IBS$$

Figure 7.7 illustrates how this cost-multiplier can be applied to existing cost estimates of IBS to estimate the cost of *Blastocystis* infection alone.

It may be worth noting that most existing studies have assumed the rate of IBS is relatively static, and that the illness is due largely to psychosomatic causes. If a substantial portion of IBS cases are caused by an infectious disease, this would suggest that the illness is contagious, and if cases are untreated, the prevalence of the illness will increase over time.

References

Al-Fellani MA, Khan AH, Al-Gazoui RM, Zaid MK, Al-Ferjani MA (2007) Prevalence and clinical features of *Blastocystis hominis* infection among patients in Sebha, Libya. Sultan Qaboos Univ Med J 7(1):35–40

Allason-Jones E, Mindel A, Sargeaunt P, Williams P (1986) *Entamoeba histolytica* as a commensal intestinal parasite in homosexual men. N Engl J Med 315(6):353–356

Amin OM (2002) Seasonal prevalence of intestinal parasites in the United States during 2000. Am J Trop Med Hyg 66(6):799–803

Bazerman MH, Moore DA (2009) Judgment in managerial decision making, 7th edn. John Wiley & Sons, Hoboken, NJ

Boorom K (2009) Emerging infectious diseases are not always obvious. Lancet Infect Dis 9(3):142

Boorom KF, Smith H, Nimri L, Viscogliosi E, Spanakos G, Parkar U, Li LH, Zhou XN, Ok UZ, Leelayoova S, Jones MS (2008) Oh my aching gut: irritable bowel syndrome, *Blastocystis*, and asymptomatic infection. Parasit Vectors 1(1):40

Batz MB, Hoffmann S, Morris JG (2012) Jr. J Food Prot. Ranking the disease burden of 14 pathogens in food sources in the United States using attribution data from outbreak investigations and expert elicitation. 75(7):1278–91

Cash B, Sullivan S, Barghout V (2005) Total costs of IBS: employer and managed care perspective. Am J Manag Care 11(suppl 1):S7–S16

Corso PS, Kramer MH, Blair KA, Addiss DG, Davis JP, Haddix AC (2003) Cost of illness in the 1993 waterborne *Cryptosporidium* outbreak, Milwaukee, Wisconsin. Emerg Infect Dis. 9(4):426–31

CDC (2000) US Center for Disease Control Fact Sheet on *Blastocystis* archived on ARCHIVE. ORG. http://web.archive.org/web/20000229131707/http://www.cdc.gov/ncidod/dpd/parasites/blastocystishominis/factsht_blastocystis_hominis.htm

CDC (2008) US Center for Disease Control Fact Sheet on *Blastocystis* archived on ARCHIVE.ORG. http://web.archive.org/web/20081014123828/http://www.cdc.gov/ncidod/dpd/parasites/blastocystishominis/factsht_blastocystis_hominis.htm

CDC (2011) US Center for Disease Control Fact Sheet on *Blastocystis* archived on ARCHIVE.ORG. http://web.archive.org/web/20100308092248/http://www.cdc.gov/ncidod/dpd/parasites/blastocystishominis/factsht_blastocystis_hominis.htm

Chandramathi S, Suresh K, Anita ZB, Kuppusamy UR (2009) Elevated levels of urinary hydrogen peroxide, advanced oxidative protein product (AOPP) and malondialdehyde in humans infected with intestinal parasites. Parasitology 136(3):359–363

Chandramathi S, Suresh K, Kuppusamy UR (2010a) Solubilized antigen of *Blastocystis hominis* facilitates the growth of human colorectal cancer cells, HCT116. Parasitol Res 106(4):941–945

Chandramathi S, Suresh K, Shuba S, Mahmood A, Kuppusamy UR (2010b) High levels of oxidative stress in rats infected with *Blastocystis hominis*. Parasitology 137(4):605–611

Chiang SF (1925) The rat as a possible carrier of the dysentery amoeba. Proc Natl Acad Sci USA 11(5):239–246

Clark CG, Diamond LS (1993) *Entamoeba histolytica*: an explanation for the reported conversion of "nonpathogenic" amebae to the "pathogenic" form. Exp Parasitol 77(4):456–460

de Lalla F, Rinaldi E, Santoro D, Nicolin R, Tramarin A (1992) Outbreak of *Entamoeba histolytica* and *Giardia lamblia* infections in travellers returning from the tropics. Infection 20(2):78–82

Dogruman-Al F, Kustimur S, Yoshikawa H, Tuncer C, Simsek Z, Tanyuksel M, Araz E, Boorom K (2009) *Blastocystis* subtypes in irritable bowel syndrome and inflammatory bowel disease in Ankara, Turkey. Mem Inst Oswaldo Cruz 104(5):724–727

Dogruman-Al F, Simsek Z, Boorom K, Ekici E, Sahin M, Tuncer C, Kustimur S, Altinbas A (2010) Comparison of methods for detection of *Blastocystis* infection in routinely submitted stool samples, and also in IBS/IBD Patients in Ankara, Turkey. PLoS One 5(11):e15484

Doyle PW, Helgason MM, Mathias RG, Proctor EM (1990) Epidemiology and pathogenicity of *Blastocystis hominis*. J Clin Microbiol 28(1):116–121

Drossman DA (2006) The functional gastrointestinal disorders and the Rome III process. Gastroenterology 130(5):1377–1390

Dyer C (2010) Lancet retracts Wakefield's MMR paper. BMJ 340(2010):c696

Editor (1999) In Memoriam Edward K. Markell. Parasitol Res 85(6):429–430. http://www.springerlink.com/content/3gvqx1tx8d8g2m7f/?MUD=MP/

El-Shazly AM, Abdel-Magied AA, El-Beshbishi SN, El-Nahas HA, Fouad MA, Monib MS (2005) *Blastocystis hominis* among symptomatic and asymptomatic individuals in Talkha Center, Dakahlia Governorate, Egypt. J Egypt Soc Parasitol 35(2):653–666

Elwakil HS, Hewedi IH (2010) Pathogenic potential of *Blastocystis hominis* in laboratory mice. Parasitol Res 107(3):685–689

Fathy FM (2011) A study on *Blastocystis hominis* in food-handlers: diagnosis and potential pathogenicity. J Egypt Soc Parasitol 41(2):433–453

Forsell J, Granlund M, Stensvold CR, Clark GC, Evengard B (2012) Subtype analysis of *Blastocystis* isolates in Swedish patients. Eur J Clin Microbiol Infect Dis 31(7):1689–1696

Giacometti A, Cirioni O, Fiorentini A, Fortuna M, Scalise G (1999) Irritable bowel syndrome in patients with *Blastocystis hominis* infection. Eur J Clin Microbiol Infect Dis 18(6):436–439

Haque R, Mondal D, Duggal P, Kabir M, Roy S, Farr BM, Sack RB, Petri WA Jr (2006) *Entamoeba histolytica* infection in children and protection from subsequent amebiasis. Infect Immun 74(2):904–909

Haque R, Mondal D, Kirkpatrick BD, Akther S, Farr BM, Sack RB, Petri WA Jr (2003) Epidemiologic and clinical characteristics of acute diarrhea with emphasis on *Entamoeba histolytica* infections in preschool children in an urban slum of Dhaka, Bangladesh. Am J Trop Med Hyg 69(4):398–405

Hussain R, Jaferi W, Zuberi S, Baqai R, Abrar N, Ahmed A, Zaman V (1997) Significantly increased IgG2 subclass antibody levels to *Blastocystis hominis* in patients with irritable bowel syndrome. Am J Trop Med Hyg 56(3):301–306

Hussein EM, Hussein AM, Eida MM, Atwa MM (2008) Pathophysiological variability of different genotypes of human *Blastocystis hominis* Egyptian isolates in experimentally infected rats. Parasitol Res 102(5):853–860

Iguchi A, Yoshikawa H, Yamada M, Kimata I, Arizono N (2009) Expression of interferon gamma and proinflammatory cytokines in the cecal mucosa of rats experimentally infected with *Blastocystis* sp. strain RN94-9. Parasitol Res 105(1):135–140

Ish-Horowicz M, Korman SH, Shapiro M, Har-Even U, Tamir I, Strauss N, Deckelbaum RJ (1989) Asymptomatic giardiasis in children. Pediatr Infect Dis J 8(11):773–779

Ithoi I, Jali A, Mak JW, Wan Sulaiman WY, Mahmud R (2011) Occurrence of *Blastocystis* in water of two rivers from recreational areas in malaysia. J Parasitol Res 2011:123916

Janoff EN, Mead PS, Mead JR, Echeverria P, Bodhidatta L, Bhaibulaya M, Sterling CR, Taylor DN (1990) Endemic *Cryptosporidium* and *Giardia lamblia* infections in a Thai orphanage. Am J Trop Med Hyg 43(3):248–256

Jimenez-Gonzalez DE, Martinez-Flores WA, Reyes-Gordillo J, Ramirez-Miranda ME, Arroyo-Escalante S, Romero-Valdovinos M, Stark D, Souza-Saldivar V, Martinez-Hernandez F, Flisser A, Olivo-Diaz A, Maravilla P (2011) *Blastocystis* infection is associated with irritable bowel syndrome in a Mexican patient population. Parasitol Res 110(3):1269–1275

Jones MS, Whipps CM, Ganac RD, Hudson NR, Boorom K (2009) Association of *Blastocystis* subtype 3 and 1 with patients from an Oregon community presenting with chronic gastrointestinal illness. Parasitol Res 104(2):341–345

Kaneda Y, Horiki N, Cheng X, Tachibana H, Tsutsumi Y (2000) Serologic response to *Blastocystis hominis* infection in asymptomatic individuals. Tokai J Exp Clin Med 25(2):51–56

Latonio AA (1988) Efficacy of a single dose of secnidazole in the treatment of acute and chronic amoebiasis. J Trop Med Hyg. 91(4):202–4

Lee MG, Rawlins SC, Didier M, DeCeulaer K (1990) Infective arthritis due to *Blastocystis hominis*. Ann Rheum Dis 49(3):192–193

Leelayoova S, Rangsin R, Taamasri P, Naaglor T, Thathaisong U, Mungthin M (2004) Evidence of waterborne transmission of *Blastocystis hominis*. Am J Trop Med Hyg 70(6):658–662

Leelayoova S, Siripattanapipong S, Thathaisong U, Naaglor T, Taamasri P, Piyaraj P, Mungthin M (2008) Drinking water: a possible source of *Blastocystis* spp. subtype 1 infection in schoolchildren of a rural community in central Thailand. Am J Trop Med Hyg 79(3):401–406

Leelayoova S, Taamasri P, Rangsin R, Naaglor T, Thathaisong U, Mungthin M (2002) In-vitro cultivation: a sensitive method for detecting *Blastocystis hominis*. Ann Trop Med Parasitol 96 (8):803–807

Li LH, Zhang XP, Lv S, Zhang L, Yoshikawa H, Wu Z, Steinmann P, Utzinger J, Tong XM, Chen SH, Zhou XN (2007a) Cross-sectional surveys and subtype classification of human *Blastocystis* isolates from four epidemiological settings in China. Parasitol Res 102(1):83–90

Li LH, Zhou XN, Du ZW, Wang XZ, Wang LB, Jiang JY, Yoshikawa H, Steinmann P, Utzinger J, Wu Z, Chen JX, Chen SH, Zhang L (2007b) Molecular epidemiology of human *Blastocystis* in a village in Yunnan province, China. Parasitol Int 56(4):281–286

Long HY, Handschack A, Konig W, Ambrosch A (2001) *Blastocystis hominis* modulates immune responses and cytokine release in colonic epithelial cells. Parasitol Res 87(12):1029–1030

Markell EK, Havens RF, Kuritsubo RA (1983) Intestinal parasitic infections in homosexual men at a San Francisco health fair. West J Med 139(2):177–178

Markell EK, Havens RF, Kuritsubo RA, Wingerd J (1984) Intestinal protozoa in homosexual men of the San Francisco Bay area: prevalence and correlates of infection. Am J Trop Med Hyg 33(2):239–245

Markell EK, Udkow MP (1986) *Blastocystis hominis*: pathogen or fellow traveler? Am J Trop Med Hyg 35(5):1023–1026

Markell EK, Udkow MP (1990a) Association of *Blastocystis hominis* with human disease? J Clin Microbiol 28(5):1085–1086

Markell EK, Udkow MP (1990b) *Blastocystis hominis*. West J Med 152(6):721

Marshall B (ed) (2002) Helicobacter pioneers: Firsthand accounts from the scientists who discovered helicobacters 1892–1982. Wiley-Blackwell ISBN 978-0867930351

Marshall BJ, Warren JR (1984) Unidentified curved bacilli in the stomach of patients with gastritis and peptic ulceration. Lancet 1(8390):1311–1315

Matukaitis JM (1997) The emerging recognition of *Cryptosporidium* as a health hazard. J Community Health Nurs 14(3):135–140

Menounos PG, Spanakos G, Tegos N, Vassalos CM, Papadopoulou C, Vakalis NC (2008) Direct detection of *Blastocystis* sp. in human faecal samples and subtype assignment using single strand conformational polymorphism and sequencing. Mol Cell Probes 22(1):24–29
Mirza H, Tan KS (2009) *Blastocystis* exhibits inter- and intra-subtype variation in cysteine protease activity. Parasitol Res 104(2):355–361
Moe KT, Singh M, Howe J, Ho LC, Tan SW, Chen XQ, Ng GC, Yap EH (1997) Experimental *Blastocystis hominis* infection in laboratory mice. Parasitol Res 83(4):319–325
Montier J (2010) The little book of behavioral investing. Wiley, New York
Nagel R, Cuttell L, Stensvold CR, Mills PC, Bielefeldt-Ohmann H, Traub RJ (2012) *Blastocystis* subtypes in symptomatic and asymptomatic family members and pets and response to therapy. Intern Med J Intern Med J. 2011 Oct 27. doi: 10.1111/j.1445-5994.2011.02626.x
Nimri L, Batchoun R (1994) Intestinal colonization of symptomatic and asymptomatic schoolchildren with *Blastocystis hominis*. J Clin Microbiol 32(11):2865–2866
Nimri LF (1993) Evidence of an epidemic of *Blastocystis hominis* infections in preschool children in northern Jordan. J Clin Microbiol 31(10):2706–2708
O'Gorman MA, Orenstein SR, Proujansky R, Wadowsky RM, Putnam PE, Kocoshis SA (1993) Prevalence and characteristics of *Blastocystis hominis* infection in children. Clin Pediatr (Phila) 32(2):91–96
Olivo-Diaz A, Romero-Valdovinos M, Gudino-Ramirez A, Reyes-Gordillo J, Jimenez-Gonzalez DE, Ramirez-Miranda ME, Martinez-Flores WA, Martinez-Hernandez F, Flisser A, Maravilla P (2012) Findings related to IL-8 and IL-10 gene polymorphisms in a Mexican patient population with irritable bowel syndrome infected with *Blastocystis*. Parasitol Res 111(1):487–491
Ortega YR, Sterling CR, Gilman RH, Cama VA, Diaz F (1993) *Cyclospora* species—a new protozoan pathogen of humans. N Engl J Med 328(18):1308–1312
Parkar U, Traub RJ, Vitali S, Elliot A, Levecke B, Robertson I, Geurden T, Steele J, Drake B, Thompson RC (2010) Molecular characterization of Blastocystis isolates from zoo animals and their animal-keepers. Vet Parasitol 169(1–2):8–17
Pickering LK, Woodward WE, DuPont HL, Sullivan P (1984) Occurrence of *Giardia lamblia* in children in day care centers. J Pediatr 104(4):522–526
Pipatsatitpong D, Rangsin R, Leelayoova S, Naaglor T, Mungthin M (2012) Incidence and risk factors of *Blastocystis* infection in an orphanage in Bangkok, Thailand. Parasit Vectors 5:37
Puthia MK, Lu J, Tan KS (2008) Blastocystis ratti contains cysteine proteases that mediate interleukin-8 response from human intestinal epithelial cells in an NF-kappaB-dependent manner. Eukaryot Cell 7(3):435–443
Qadri F, Saha A, Ahmed T, Al Tarique A, Begum YA, Svennerholm AM (2007) Disease burden due to enterotoxigenic *Escherichia coli* in the first 2 years of life in an urban community in Bangladesh. Infect Immun 75(8):3961–3968
Qadri SM, al-Okaili GA, al-Dayel F (1989) Clinical significance of *Blastocystis hominis*. J Clin Microbiol 27(11):2407–2409
Rendtorff RC (1954) The experimental transmission of human intestinal protozoan parasites. II. *Giardia lamblia* cysts given in capsules. Am J Hyg 59(2):209–220
Rendtorff RC, Holt CJ (1954) The experimental transmission of human intestinal protozoan parasites. IV. Attempts to transmit *Endamoeba coli* and *Giardia lamblia* cysts by water. Am J Hyg 60(3):327–338
Rosenblatt JE (1990) *Blastocystis hominis*. J Clin Microbiol 28(10):2379–2380
Scicluna SM, Tawari B, Clark CG (2006) DNA barcoding of *Blastocystis*. Protist 157(1):77–85
Senay H, MacPherson D (1990) *Blastocystis hominis*: epidemiology and natural history. J Infect Dis 162(4):987–90
Silverman RH, Das Gupta J, Lombardi VC, Ruscetti FW, Pfost MA, Hagen KS, Peterson DL, Ruscetti SK, Bagni RK, Petrow-Sadowski C, Gold B, Dean M, Mikovits JA (2011) Partial retraction. Detection of an infectious retrovirus, XMRV, in blood cells of patients with chronic fatigue syndrome. Science 334(6053):176
Smith JW, Wolfe MS (1980) Giardiasis. Annu Rev Med 31:373–383

Souppart L, Moussa H, Cian A, Sanciu G, Poirier P, El Alaoui H, Delbac F, Boorom K, Delhaes L, Dei-Cas E, Viscogliosi E (2010) Subtype analysis of *Blastocystis* isolates from symptomatic patients in Egypt. Parasitol Res 106(2):505–511

Souppart L, Sanciu G, Cian A, Wawrzyniak I, Delbac F, Capron M, Dei-Cas E, Boorom K, Delhaes L, Viscogliosi E (2009) Molecular epidemiology of human *Blastocystis* isolates in France. Parasitol Res 105(2):413–421

Stein J (2007) Drug combination promising for enteric parasitic infections. Reuters Health (Indexed in Medcape) (ID 565247)

Stensvold CR, Arendrup MC, Jespersgaard C, Molbak K, Nielsen HV (2007) Detecting *Blastocystis* using parasitologic and DNA-based methods: a comparative study. Diagn Microbiol Infect Dis 59(3):303–307

Stensvold CR, Christiansen DB, Olsen KE, Nielsen HV (2011) *Blastocystis* sp. subtype 4 is common in Danish *Blastocystis*-positive patients presenting with acute diarrhea. Am J Trop Med Hyg 84(6):883–885

Stensvold CR, Lewis HC, Hammerum AM, Porsbo LJ, Nielsen SS, Olsen KE, Arendrup MC, Nielsen HV, Molbak K (2009) *Blastocystis*: unravelling potential risk factors and clinical significance of a common but neglected parasite. Epidemiol Infect 137(11):1655–1663

Stensvold CR, Smith HV, Nagel R, Olsen KE, Traub RJ (2010) Eradication of *Blastocystis* carriage with antimicrobials: reality or delusion? J Clin Gastroenterol 44(2):85–90

Stensvold R, Brillowska-Dabrowska A, Nielsen HV, Arendrup MC (2006) Detection of *Blastocystis hominis* in unpreserved stool specimens by using polymerase chain reaction. J Parasitol 92(5):1081–1087

Stolt CM, Klein G, Jansson AT (2004) An analysis of a wrong Nobel Prize-Johannes Fibiger, 1926: a study in the Nobel archives. Adv Cancer Res 92:1–12

Surangsrirat S, Thamrongwittawatpong L, Piyaniran W, Naaglor T, Khoprasert C, Taamasri P, Mungthin M, Leelayoova S (2010) Assessment of the association between *Blastocystis* infection and irritable bowel syndrome. J Med Assoc Thai 93(Suppl 6):S119–S124

Suresh K, Smith H (2004) Comparison of methods for detecting *Blastocystis hominis*. Eur J Clin Microbiol Infect Dis 23(6):509–511

Suresh K, Smith HV, Tan TC (2005) Viable *Blastocystis* cysts in Scottish and Malaysian sewage samples. Appl Environ Microbiol 71(9):5619–5620

Tan TC, Suresh KG (2006a) Amoeboid form of *Blastocystis hominis*—a detailed ultrastructural insight. Parasitol Res 99(6):737–742

Tan TC, Suresh KG (2006b) Predominance of amoeboid forms of *Blastocystis hominis* in isolates from symptomatic patients. Parasitol Res 98(3):189–193

Tungtrongchitr A, Manatsathit S, Kositchaiwat C, Ongrotchanakun J, Munkong N, Chinabutr P, Leelakusolvong S, Chaicumpa W (2004) *Blastocystis hominis* infection in irritable bowel syndrome patients. Southeast Asian J Trop Med Public Health 35(3):705–710

Udkow MP, Markell EK (1993) *Blastocystis hominis*: prevalence in asymptomatic versus symptomatic hosts. J Infect Dis 168(1):242–244

US EPA (1979) Proceedings of the EPA symposium on giardiasis. http://nepis.epa.gov/Exe/ZyPURL.cgi?Dockey=2000UCBL.txt.179

Utzinger J, Botero-Kleiven S, Castelli F, Chiodini PL, Edwards H, Kohler N, Gulletta M, Lebbad M, Manser M, Matthys B, N'Goran EK, Tannich E, Vounatsou P, Marti H (2010) Microscopic diagnosis of sodium acetate-acetic acid-formalin-fixed stool samples for helminths and intestinal protozoa: a comparison among European reference laboratories. Clin Microbiol Infect 16(3):267–273

Veenemans J, Mank T, Ottenhof M, Baidjoe A, Mbugi EV, Demir AY, Wielders JP, Savelkoul HF, Verhoef H (2011) Protection against diarrhea associated with *Giardia intestinalis* is lost with multi-nutrient supplementation: a study in Tanzanian children. PLoS Negl Trop Dis 5(6):e1158

Wells JG, Davis BR, Wachsmuth IK, Riley LW, Remis RS, Sokolow R, Morris GK (1983) Laboratory investigation of hemorrhagic colitis outbreaks associated with a rare *Escherichia coli* serotype. J Clin Microbiol 18(3):512–520

Whipps CM, Boorom K, Bermudez LE, Kent ML (2010) Molecular characterization of *Blastocystis* species in Oregon identifies multiple subtypes. Parasitol Res 106(4):827–832

Windsor JJ (2007) *B. hominis* and *D. fragilis*: neglected human protozoa. The Biomedical Scientist, pp. 524–527

Wyatt RG, Dolin R, Blacklow NR, DuPont HL, Buscho RF, Thornhill TS, Kapikian AZ, Chanock RM (1974) Comparison of three agents of acute infectious nonbacterial gastroenteritis by cross-challenge in volunteers. J Infect Dis 129(6):709–714

Yakoob J, Jafri W, Beg MA, Abbas Z, Naz S, Islam M, Khan R (2010a) *Blastocystis hominis* and *Dientamoeba fragilis* in patients fulfilling irritable bowel syndrome criteria. Parasitol Res 107(3):679–684

Yakoob J, Jafri W, Beg MA, Abbas Z, Naz S, Islam M, Khan R (2010b) Irritable bowel syndrome: is it associated with genotypes of *Blastocystis hominis*. Parasitol Res 106(5):1033–1038

Yakoob J, Jafri W, Jafri N, Islam M, Asim Beg M (2004a) In vitro susceptibility of *Blastocystis hominis* isolated from patients with irritable bowel syndrome. Br J Biomed Sci 61(2):75–77

Yakoob J, Jafri W, Jafri N, Khan R, Islam M, Beg MA, Zaman V (2004b) Irritable bowel syndrome: in search of an etiology: role of *Blastocystis hominis*. Am J Trop Med Hyg 70(4):383–385

Yao FR, Qiao JY, Zhao Y, Zhang X, Yang JH, Li XQ (2005) Experimental infection of mice with *Blastocystis hominis*. Zhongguo Ji Sheng Chong Xue Yu Ji Sheng Chong Bing Za Zhi 23 (6):444–448

Yoshikawa H, Wu Z, Kimata I, Iseki M, Ali IK, Hossain MB, Zaman V, Haque R, Takahashi Y (2004) Please add these references: Polymerase chain reaction-based genotype classification among human *Blastocystis hominis* populations isolated from different countries. Parasitol Res. 92(1):22–9. Epub 2003 Nov 4

Zaman V, Zaki M (1996) Resistance of *Blastocystis hominis* cysts to metronidazole. Trop Med Int Health 1(5):677–678

Zhang HW, Li W, Yan QY, He LJ, Su YP (2006) Impact of *Blastocystis hominis* infection on ultrastructure of intestinal mucosa in mice. Zhongguo Ji Sheng Chong Xue Yu Ji Sheng Chong Bing Za Zhi 24(3):187–191

Zierdt CH (1991) Pathogenicity of *Blastocystis hominis*. J Clin Microbiol 29(3):662–663

Chapter 8
Behavioral Decision Analysis and Pathogenicity: How Do We Decide What Makes Us Sick?

Kenneth Boorom

Abstract The previous chapter suggested ways to use statistics to understand conflicting study results and improve decision making, however, such approaches remain uncommon in the area of infectious diseases. This chapter examines the processes that are currently used to resolve controversies, and the criteria people use to judge pathogenicity. We examine methods by which a scientific controversy can be categorized and understood based on written statements by individuals involved. Existing published studies suggest that the idea of pathogenicity is strongly subjective, with significance variation from individual to individual. Considerations such as economic cost, perceived benefit, and conflict with religious and philosophical beliefs can influence researcher viewpoints. By understanding the particular interests and impacts such a decision makes on different interest groups, scientists can design studies and guide research efforts to avoid unproductive conflict with the medical community.

Keywords *Blastocystis* • Pathogenicity • Irritable bowel syndrome • Diarrheal diseases

8.1 Introduction

The last chapter focused on statistical ways to analyze medical literature to obtain conclusions, but in practice most decisions people make are not based on statistical analysis (Bazerman and Moore 2009). This chapter investigates the opinions people

Electronic supplementary material: The online version of this article (doi: 10.1007/978-3-642-32738-4) contains supplementary material, which is available to authorized users.

K. Boorom (✉)
Blastocystis Research Foundation, 5060 SW Philomath Blvd, #202, Corvallis, OR 97333, USA
e-mail: kboorom@bhomcenter.org

H. Mehlhorn et al. (eds.), *Blastocystis: Pathogen or Passenger?*,
Parasitology Research Monographs 4,
DOI 10.1007/978-3-642-32738-4_8,

have formed about *Blastocystis*, and the factors that may have influenced those beliefs. The documents used to form this view consist of the following:

1. Medical studies and letters concerning *Blastocystis* published in peer reviewed journals.
2. Informational sheets and "fact sheets" on *Blastocystis* published by health organizations.
3. Correspondence from the US National Institutes of Health (NIH) and US Center for Disease Control to US Congressional Representatives and BRF concerning *Blastocystis* infection.
4. Internal correspondence between employees of the NIH concerning *Blastocystis* infection, which BRF obtained through the Freedom of Information Act (FOIA).
5. Public testimony provided by the State of Oregon's Epidemiology Office at a hearing of legislation to make *Blastocystis* a reportable infection in Oregon.
6. Funding history for research into *Blastocystis*, from 1985 to 2012, as obtained from the NIH's REPORTer database.
7. Journal articles and text books analyzing the phenomenon of scientific controversy.
8. Emails and letters from patients and physicians to BRF concerning *Blastocystis* infection.

Although many of these sources are from the USA, it is suggested the behavior is of general interest, because most of the published opposition to *Blastocystis* as a pathogen has come from the USA and other English-speaking countries (Chapter 7, Table 1.3). Most of the world's most influential medical journals (JAMA, Lancet, NEJM, Annals of Internal Medicine) are published in the USA or the UK, raising the possibility that medical opinion in those countries may have a disproportionate impact on global medical thought. Finally, the influence of medical information available through the internet is increasing in society, and the highest ranked documents concerning *Blastocystis* infection are commonly authored in the USA by the US Center for Disease Control and the Mayo Clinic, according to Google's search ranking results.

In its simplest form, assessing pathogenicity involves determining whether exposure to a microbe will result in illness in some patients who would not be sick if they were not exposed. But analysis of the process as a purely scientific endeavor can't account for the delays of decades between the first reports of illness associated with a microbe, and a consensus within the medical community that it should be classified as pathogenic (Table 7.8)

Over 15 years ago, researchers were already noting the slow rate of progress in this area:

> Despite more than 80 years of debate since the first accepted descriptions of the genus *Blastocystis*, not a single issue about this organism has been satisfactorily resolved. Substantial progress has been made in the description of the morphology of *Blastocystis hominis* during recent years, but structures and organelles present in the cell remain of unknown function. The method(s) of division is questioned. The pathogenicity of the organism has not been determined, and the significance of *B. hominis* in immunocompromised patients has not been ascertained. The necessity for treatment and the most appropriate chemotherapeutic strategies have not been defined. The taxonomy of the organism remains controversial.
>
> *Blastocystis Revisited* (Stenzel and Boreham 1996)

Considering the progress commonly seen in "hard sciences" (atomic physics, astronomy, electronics, aeronautics) over 80 years, it is difficult to attribute the delay associated with assessment of pathogenicity as stemming from purely scientific or technical barriers.

8.2 Cognitive Controversy Vs. Social Controversy

Disagreement about the significance of scientific data is not unique to *Blastocystis*, and sociologists have produced a body of literature that classifies controversies, and describes the interactions that occur in their formation and resolution. An awareness of concepts from this area of study can be valuable to researchers seeking to understand what may be seemingly irrational behaviors in the field of medical science.

Sociologists suggest that scientific controversies can be divided into two classes to facilitate analysis (Engelhardt and Caplan 1987). *Cognitive controversies* concern specific facts or scientific findings, while *social controversies* involve decisions about public policy. An example of a purely cognitive controversy might be the finding, reported by researchers in 2011, that neutrinos could be observed to travel faster than light (Cartlidge 2012). Because the outcome of the controversy would have little or no impact on any social policy, the controversy depended solely on the factual question as to whether neutrinos could travel at the proposed speed. A purely social controversy would be the disagreement about public funding of birth control information in predominantly Catholic countries. Because an abundance of information exists on the clinical efficacy and use of birth control methods, there are few factual questions that would be relevant to the resolution of this disagreement. Disagreements appear to stem solely from personal and religious beliefs.

Social controversies commonly involve a larger number of people, many of whom are not experts in a particular field, and who are participating due to an economic or personal interest in the debate's outcome. Social controversies can continue for decades or generations, because they involve competing interests and philosophical disagreements which are not resolved by the presence of new information. Cognitive controversies are often resolved in a few years. For example, the issue of faster-than-light neutrinos was explained in a few months once other groups tried to repeat the findings (Cartlidge 2012).

Which model best describes the controversies associated with assessment of pathogenicity? Although this may seem like a simple cognitive question, the timing of events suggests that the ability to demonstrate the pathogenicity of most enteric microbes predated the acceptance of pathogenicity by decades. For example, references to *Giardia intestinalis* pathogenicity date back to World War I, with a 1916 paper (Fantham and Porter 1916) noting that "While British troops were in Gallipoli, a number of men contracted various forms of dysentery or diarrhea, and severe cases occurred. Among the men invalided to England a number were found to be infected with Giardia (Lamblia)." In addition to reporting the illness, the authors of the paper performed an experimental animal infection on rats and kittens:

> From then till the forty-fourth day the feces were negative. On the forty-fourth day the feces were loose and diarrheic, and *Lamblia* cysts and flagellates appeared in them. The kitten showed signs of distress, howled, and refused food. Its coat became rough, and the animal shivered. On the forty-fifth, forty-seventh, and forty-eighth days some *Lamblia* cysts were present in the feces. The kitten seemed ill, vomited, sent up piteous cries, and the condition of the coat was bad. Violent diarrhea set in on the forty-eighth day, and the kitten died on the forty-ninth day after the first infective feed. At death it weighed 638 grams, its control weighing 997 grams.
>
> *The Pathogenicity of Giardia (Lamblia) Intestinalis to Men and to Experimental Animals* (Fantham and Porter 1916)

In the case of *G. intestinalis*, we can see that the technological means exist to demonstrate pathogenicity as early as 1916, but 40 years later, medical professionals still rejected the idea that it could cause epidemic illness. A paper reporting an epidemic of giardiasis in Portland, Oregon impacting over 50,000 individuals was rejected in the mid-1950s on the grounds that demonstration of *G. intestinalis* infection in a large portion of the population did not constitute proof of an epidemic (US_EPA 1979). Oregon's public health commissioner attributed the illness to an unknown virus and is quoted as such in the proceedings of the 1978 EPA Symposium on Waterborne Transmission of Giardia (US_EPA 1979). Those proceedings are now available online from the National Service Center for Environmental Publications, and they provide a fascinating look at the struggles scientists had in establishing *G. intestinalis* as a cause of disease (US_EPA 1979).

8.3 Resolution of Scientific Controversies

Most discussion of *Blastocystis* controversy has focused on specific findings concerning that infection. But the subject of *Blastocystis* controversy can be viewed in a larger context as just one of a number of scientific or medical controversies which can be found to exist at any point in history. In a text devoted to the subject, medical ethicist H. Tristam Engelhardt Jr. (Center for Ethics, Medicine, and Public Issues, Baylor College of Medicine) identifies five ways in which scientific controversies can be resolved (Engelhardt and Caplan 1987):

1. *Sound argument*: An argument is presented that is accepted by all parties and shows that the opposing view is incorrect.
2. *Consensus closure*: The opposing views have not been found to be incorrect, but the parties agree to end the controversy.
3. *Procedural closure*: The controversy is ended by a formal, governed effort which ends the discussion.
4. *Natural death closure*: The controversy has ended by the gradual fading away of interest.
5. *Negotiation closure*: The controversy has ended because a negotiated settlement has been reached which is acceptable to the opposing parties.

In the last few years, a number of cognitive controversies in science have been resolved when other scientists have tried to repeat an experiment, failed, and sometimes succeeded in identifying an intervening variable that was present in the original research report, and lead to a faulty conclusion. For example, the phenomenon of cold fusion was not reproducible outside of the laboratories where it was discovered and may have been the result of measurement error (Huizenga 1992). When researchers were unable to reproduce the results linking xenotropic murine leukemia virus-related virus (XMRV) virus with Chronic Fatigue Syndrome (CFS), analysis by other labs showed that commercially available PCR reagents were commonly contaminated with the virus, which could have led to the original researcher's unreproducible findings (Zheng et al. 2011).

However, in most of the controversies surrounding gastrointestinal microbes, the resolution of the controversy did not coincide with a particular study or technique, or any action by the scientific community. Scientists had developed a model for experimental giardiasis as early as 1916 (Fantham and Porter 1916), and *Cryptosporidium spp.* was repeatedly identified as an agent of diarrhea in immunocompetent individuals in the mid-1980s (O'Donoghue 1985). During the 1980s, and 1990s, when cases were emerging by what could now be estimated in the millions (Kappus et al. 1994), scientific and medical organizations were largely silent on the issue of public response to *G. intestinalis* and *Cryptosporidium spp.* These included prestigious groups such as the American Medical Association, the National Academy of Sciences, the American Society of Microbiology, the US Center for Disease Control, the US NIH, and the Infectious Disease Society of America (IDSA).

In fact, it was the US EPA which took action unilaterally to organize researchers to identify and develop a response to highly publicized epidemics from drinking water contamination (Eisenberg et al. 2005; Matukaitis 1997; Smith and Wolfe 1980; US_EPA 1979). Actions by that organization, rather than a scientific or medical society, are credited by some researchers with recognition of the significance of the problem of *G. intestinalis* infection (Smith and Wolfe 1980). Giardiasis finally became a nationally reportable illness in the USA in 2002, 87 years after the 1916 publication of Fantham's animal model for the infection (Hlavsa et al. 2005).

8.4 Role of the NIH in *Blastocystis* Controversy

The US NIH is the world's largest funding body for medical research. In 2011, the agency's budget was $31 billion, which was distributed between internal research projects and over 40,000 external investigators (Crow 2011; Ginther et al. 2011). The NIH played an intriguing role in the controversy surrounding *Blastocystis*, in that the agency had a role in beginning the controversy, by funding a large part of the research into the organism in the 1980s. Afterwards, the NIH actions contributed to an end of research in the US, by making statements that discouraged researchers from investigating the infection.

Between 1976 and 1995, the NIH was the world's leading institution concerning research on *Blastocystis* infection (Table 8.1). All papers originating from the NIH labs published after 1986 identified *Blastocystis* as pathogenic (Table 8.1). NIH researcher Dr. Charles H. Zierdt authored several of the most frequently referenced paper of *Blastocystis* infection, including the second most frequently referenced review paper (Zierdt 1991a). He also contributed the first study to demonstrate that *Blastocystis* patients exhibited an elevated serum antibody response to *Blastocystis* antigen not seen in healthy controls (Zierdt et al. 1995). Dr. Zierdt performed many of the early studies which developed ways to culture the organism, as well as axenicizing isolates from patients into the collection of the American Tissue and Culture Collection (ATCC) (Jones 2008).[1]

Interestingly, the NIH appears to have reversed its position on funding of *Blastocystis* research after the 1990s. Communications from researchers and patients interacting with the agency indicate that the agency considered *Blastocystis* to be nonpathogenic, and ineligible for funding:

> When I've contacted the NIH about *Blastocystis* research, they've told me there is no evidence that *Blastocystis* can cause illness in humans.
> US Scientist to BRF, 2010 (name withheld)

The duality of the NIH's response is even more intriguing, as the agency's own Medical Terms Database (MESH) clearly identifies *Blastocystis* as a pathogen (Fig. 8.1). Interestingly. the entry also describes disabling illness, which is consistent with descriptions of illness provided by some patients, but had not been published widely in scientific literature.

8.4.1 *Obtaining Data from the NIH: Methods and Results*

The NIH's position on *Blastocystis* research appears to be a complex one, and one that varies with time. Understanding the agency's position further is a complex task, since the NIH does not publish planning or road-mapping documents for most of the infectious diseases studied at that agency, so one cannot point to a date where a specific planning document changed.

The NIH's grant award database provided one source of information. The NIH Reform Act, passed by the US Congress in 2006, required the agency to make its grant awards database accessible by the public. Queries to that grant database show that between 1995 and May 2012, the NIH approved a total of over 600 grants into *G. intestinalis*, *Cryptosporidium spp.*, *and Entamoeba histolytica* with a cost of $100 million, but 0 grants were approved for *Blastocystis* research (Table 8.2).

---

[1] Cultures axenicized by Dr. Zierdt are still available from the ATCC and were used to develop the first real-time PCR test, which was used in the first US study to demonstrate that patients with symptoms attributable to irritable bowel syndrome or Gulf War Illness were infected with *Blastocystis*, which had not been previously detected in those patients (Jones 2008, 2009).

Table 8.1 NIH Papers on *Blastocystis* Infection

N	Date	Title	Journal
1	Oct 1995	Serum antibody detected by fluorescent antibody test in patients with symptomatic *Blastocystis hominis* infection	Recenti Prog Med
2	Feb 1995	Enzyme-linked immunosorbent assay for detection of serum antibody to *Blastocystis hominis* in symptomatic infections	Journal of Parasitology
3	Apr 1994	Lipid biosynthesis by axenic strains of *Blastocystis hominis*	Comp Biochem Physiol Biochem Mol Biol
4	Jun 1993	Antibody response to *Blastocystis hominis* infections	Annals of Internal Medicine
5	Jan 1993	Taxonomic status of *Blastocystis hominis*: reply	Parasitology Today
6	Jul 1992	Comparative analysis of lipid composition in axenic strains of *Blastocystis hominis*	Comp Biochem Physiol B
7	Mar 1991	Pathogenicity of *Blastocystis hominis*	Journal of Clinical Microbiology
8	Jan 1991	*Blastocystis hominis*—past and future	Clinical Microbiology Rev
9	Sep 1990	Magainin analogs effective against pathogenic protozoa	Antimicrobial Agents and Chemotherapy
10	May 1988	Biochemical and ultrastructural study of *Blastocystis hominis*	Journal of Clinical Microbiology
11	Jan 1988	*Blastocystis hominis*, a long-misunderstood intestinal parasite	Parasitology Today
12	Feb 1986	Cytochrome-free mitochondria of an anaerobic protozoan—*Blastocystis hominis*	Journal of Protozoology
13	May 1983	In vitro response of *Blastocystis hominis* to antiprotozoal drugs	Journal of Protozoology
14	Nov 1981	Generation time and growth rate of the human intestinal parasite *Blastocystis hominis*	Journal of Protozoology
15	Oct 1976	Ultrastructure and light microscope appearance of *Blastocystis hominis* in a patient with enteric disease	Zeitschrift für Parasitenkunde
16	Jun 1976	Endosymbiosis in *Blastocystis hominis*	Experimental Parasitology
17	Jun 1976	*Blastocystis hominis*: pathogenic potential in human patients and in gnotobiotes	Experimental Parasitology
18	Oct 1974	*Blastocystis hominis*: axenic cultivation	Experimental Parasitology
19	1974	Freeze-etch studies of the granular and vacuolated forms of *Blastocystis hominis*	Zeitschrift für Parasitenkunde
20	Nov 1973	Ultrastructure of *Blastocystis hominis*	Zeitschrift fur Parasitenkunde
21	Feb 1973	Studies of *Blastocystis hominis*	Journal of Protozoology
22	Nov 1967	Protozoan characteristics of *Blastocystis hominis*	American Journal of Clinical Pathology

The US National Institutes of Health was a major contributor to *Blastocystis* research until the mid-1990s, with over 22 articles published by Dr. Charles H Zierdt on the subject. All of the Zierdt's later papers identified *Blastocystis* as pathogenic. Z. Parasitenkd; now Parasitology Research

National Library of Medicine - Medical Subject Headings

2010 MeSH

MeSH Descriptor Data

Return to Entry Page

Standard View. Go to Concept View; Go to Expanded Concept View

MeSH Heading	Blastocystis hominis
Tree Number	B01.046.500.100.200.200.375
Annotation	infection: coordinate IM with BLASTOCYSTIS INFECTIONS (IM); coordinate with specific protozoan terms (ANTIGENS, PROTOZOAN, etc) if pertinent
Scope Note	A species of parasitic protozoa found in the intestines of humans and other primates. It was classified as a yeast in 1912. Over the years, questions arose about this designation. In 1967, many physiological and morphological B. hominis characteristics were reported that fit a protozoan classification. Since that time, other papers have corroborated this work and the organism is now recognized as a protozoan parasite of humans causing intestinal disease with potentially disabling symptoms.
Allowable Qualifiers	CH CL CY DE EN GD GE IM IP ME MI PH PS PY RE UL VI
Previous Indexing	Phycomyces (1971-1979)
Previous Indexing	Protozoa (1986-1991)
Previous Indexing	Zygomycotina (1980-1985)
History Note	92
Date of Entry	19910502
Unique ID	D016845

MeSH Tree Structures

Fig. 8.1 Entry for *Blastocystis* infection in the NIH's MESH Database, Retrieved July 2011. The entry illustrates the conflicting positions from that agency. Interestingly, the entry makes note of "disabling" symptoms, which have been frequently described to BRF in patient emails, but are generally undocumented in medical literature

The grant awards database made it possible to document an outcome (funding behavior), but it did not provide visibility into the thought process responsible for producing that outcome. Letters from BRF and US Congressman generated written responses from the agency. A 2008 response indicated *Blastocystis* was a special case, noting that a commitment of funding for clinical research was "infeasible," but researchers could try applying for grants (Fig. 8.1 and Supplemental files).

From 2009 to 2011, the NIH's position on the infeasibility of performing clinical *Blastocystis* research did not change, despite the emergence of a body of literature

Table 8.2 Comparison of the number of NIH grants and aggregate spending for selected enteric microbes from January 1995–May 2012

Organism	Number of NIH grants written with microbe term in title 1995–2012	Total dollar amount of grants (sum of primary project dollar column)	Frequency of detection in symptomatic patients based on 2000 US-wide survey (Amin 2002)	Percent of mono-infections that were symptomatic based on 2000 US-wide survey (Amin 2002)
Blastocystis sp.	0	$0	24 % (662/2,896)	69 %
Giardia intestinalis	176	$31,812,247	0.4 % (19/2,896)	100 %
Cryptosporidium spp.	267	$44,954,253	4.1 % (121/2,896)	70 %
Entamoeba histolytica	180	$24,080.646	2.3 % (68/2,896)	75 %

published internationally, which became progressive more detailed about how it was causing illness, and how to develop animal models. As international research groups had clearly made a different decision about *Blastocystis* research, we wanted to understand why the NIH's decision differed. Mainly, why would different scientific organizations draw very different conclusions, given the same body of research study available? What were the criteria by which the NIH had determined should be treated specially? Why was the funding level remaining at zero for so many years in a row? And why did researchers indicate that the NIH had advised them against applying for grants?

In 2011, BRF filed a FOIA request with the agency to obtain documents relating to the decision making process concerning *Blastocystis*.[2] The FOIA request asked for information similar to that requested in a November, 2011 letter from US Congressman Kurt Schrader to the NIH Supplementary files. The NIH responded to that letter, but the response did not answer the questions posed (Supplementary files). The FOIA request specified the following documents:

1. Written documents which describe the process by which NIAID reviews medical literature to assess the public health significance of specific microbes.
2. Documents from meetings held in 2008, 2009, 2010, and 2011 where the NIAID reviewed the public health significance of *Blastocystis* and/or its policies and research activities related to *Blastocystis* infection. The request included emails, agendas, meeting notes, and handouts.

[2] The intention in filing the FOIA request was not to be disrespectful of the institution or its staff members. Rather, BRF noted that each US University-based research effort that we had worked to organized came to an end when it was determined that the NIH would not provide funding. At the same time, we saw groups emerging in Mexico, Europe, and Asia, many of which did receive national funding. Why was the NIH's position different? What can we learn about the decision making process from the differences? We also felt that this was a legitimate public policy question, since this infection was present in 10–15 % of the US population, with about 70 % of mono-infections being symptomatic by some analyses (Amin 2002).

DEPARTMENT OF HEALTH & HUMAN SERVICES

Public Health Service

National Institutes of Health
National Institute of Allergy
and Infectious Diseases
Bethesda, Maryland 20892

May 28, 2008

Mr. Ken Boorom
Director, Blastocystis Research Foundation
5060 SW Philomath Blvd., #202
Corvallis, OR 97333-1044

Dear Mr. Boorom:

Thank you for your letter of May 12, 2008, to Dr. Anthony S. Fauci, Director of the National Institute of Allergy and Infectious Diseases (NIAID), a component of the National Institutes of Health (NIH), concerning funding for the development of diagnostics and treatments for *Blastocystis hominis* infection. Thank you also for your note transmitting a copy of the letter to Dr. Elias Zerhouni, Director of the NIH. As the Principal Deputy Director of NIAID, I am pleased to respond to your letter.

NIAID is committed to funding innovative basic research, as well as the development and clinical testing of vaccines, diagnostics, and therapeutics for a wide variety of infectious and immunologic diseases, including diseases caused by parasites and protozoans. Because many questions remain regarding the ability of *B. hominis* to cause disease, more basic research needs to be done before the research that you suggest would be feasible. Although NIAID currently is not funding research directly related to *B. hominis*, should a researcher submit a grant application to NIH regarding *B. hominis*, it would be reviewed, like all grant applications, through the NIH peer review process, which is designed to evaluate and rate the scientific and technical merit of research applications for possible consideration for funding. More detailed information on the NIH peer review process can be found at the following Web site: http://grants1.nih.gov/grants/peer/peer.htm.

Thank you for your interest in NIAID research on *B. hominis*, and for your leadership and advocacy on behalf of other patients. I hope that this information is helpful to you. Please do not hesitate to contact me if I can be of further assistance.

Sincerely,

Hugh Auchincloss, M.D.
Principal Deputy Director
National Institute of Allergy and
Infectious Diseases

Fig. 8.2 Letter from the NIH indicating that it was not pursuing any *Blastocystis* research, but that it would consider grants

3. Responses by the Laboratory of Parasitic Diseases to inquiries outside of the NIAID into whether *Blastocystis* should be considered a pathogen, and into any aspect concerning the scientific research or clinical status or *Blastocystis* infection including email responses and written responses to physicians, public health professionals, researchers, and patients.

The FOIA request produced only two email communications (Figs. 8.3, 8.4, 8.5). We found no evidence of any organized discussion or review of *Blastocystis* literature, or of any meetings conducted from 2008 to 2011 to discuss the issue of *Blastocystis* funding or pathogenicity (Table 8.3).

Schmidt, Beth (NIH/NIAID) [E]

Subject: FW: AFC referral- blastocystis

From: Tom Nutman <tnutman@niaid.nih.gov>
Date: Mon, 22 Nov 2010 16:24:11 -0500
To: JeanAnne Ware <jeananne.ware@nih.gov>
Cc: Kathryn Spates <spatesk@mail.nih.gov>
Subject: Re: AFC referral- blastocystis

I think you can pass this along to Kate tomorrow. We will likely not see the patient in that they have 2 non-pathogens and no eosinophilia. If they feel they must treat the B. hominis - now considered a fungus and not a parasite -- they should consider using metronidizole 750 mg TID x 10 d or Trimethoprim/sulfa 1 DS po BID x 7d. We would not treat this patient, however. Tom

On 11/22/10 4:20 PM, "JeanAnne Ware" <jeananne.ware@nih.gov> wrote:

Dr. Nutman,

We received some lab results on a patient from AFC whom they are referring for blastocystis hominis in the stool, trichrome stain showing few b. hominis, dientamoeba fragilis (there is just one O &P result). Aside from that, the patient does not have eosinophilia and we not have any other clinical information besides labs. Would this patient be appropriate for screening, or is there this anything else you'd like me to do with this, i.e. contact the provider with any guidance, or pass this along to Kate for any additional screening? (she is gone for the day so I am picking up the incoming labs).

JeanAnne
JeanAnne Ware, CRNP
Nurse Practitioner
Laboratory of Parasitic Diseases
National Institute of Allergy and Infectious Diseases
10 Center Drive, Building 10, Room 8N234-A
Bethesda, MD 20892
Phone: 301.594.3149
Fax: 301.451.2029
Pager: 102-10605
Email: warejm@niaid.nih.gov

The information in this e-mail and any of its attachments is confidential and may contain sensitive information. It should not be used by anyone who is not the original intended recipient. If you have received this e-mail in error please inform the sender and delete it from your mailbox or any other storage devices. National Institute of Allergy and Infectious Diseases shall not accept liability for any statement made that are sender's own and not expressly made on behalf of the NIAID by one of its representatives.

--
Thomas B. Nutman, M.D.
Head, Helminth Immunology Section and
Head, Clinical Parasitology Unit
Laboratory of Parasitic Diseases
Bldg 4 Room B1-03
4 Center Dr.
National Institutes of Health
Bethesda, MD 20892-0425

Fig. 8.3 Decision as to whether the NIH Laboratory of Parasitic Diseases would consider seeing a patient with *Blastocystis* spp. infection as well as *Dientamoeba fragilis* infection. The response from the Deputy Chief indicates that the laboratory considers both to be nonpathogenic. The response notes that *Blastocystis* has been reclassified as a fungus, and is no longer a parasite

From: Allen, Andrew [mailto:aallen@jcvi.org]
Sent: Tuesday, June 08, 2010 1:48 AM
To: Giovanni, Maria (NIH/NIAID) [E]; Wang, Lu (NIH/NHGRI) [E]
Subject: Blastocystis - biomedical relevance

I just wanted to touch base with you again concerning a possible Blastocystis project.

I have spoken to many people in the scientific community who view this as an extremely high priority project. This includes a community of domestic and international physicians who continue to be extremely frustrated by lack of treatment as well as prominent scientists within the pathogen genomics community that readily acknowledge a major knowledge gap when it comes to Blastocystis. Blastocystis is widely recognized by the international parasite research community as a highly problematic emerging pathogen. This evident from many conversations I have had with different researchers as well as an objective surrey of the recent literature.

Blastocystis is classified as a pathogen by the CDC. In fact, Blastocystis is the single most common eukaryotic parasite in the intestinal tract of humans. Blastocystis infection is having a very significant impact on public health world-wide. There is highly significant biomedical relevance for development of genome-enabled approaches to the study Blastocystis. Long term Blastocystis infection is common and methods for diagnosis and treatment remain rudimentary at best.

Considering the biomedical importance Blastocystis and the existing knowledge-gap, I wondered if there is anything I can do in order to attempt to initiate a GSC project on Blastocystis.

Thanks very much.

Sincerely, Andy

Andrew E. Allen, Ph.D.
Associate Professor
J. Craig Venter Institute
Dept. of Microbial and Environmental Genomics
10355 Science Center Drive
San Diego, CA 92121
Phone: 858-200-1826
Fax: 858-200-1881

Fig. 8.4 An inquiry from a US researcher to the NIH about the possibility of funding a research project into Blastocystis infection. Internal discussion of the inquiry at the NIH appears in Fig. 8.5

The FOIA documents more closely reflected the information we received from researchers. Those emails indicated that staff members considered the question of whether the infection was pathogenic to be resolved in favor of nonpathogenicity, and did not wish to revisit the issue (Figs. 8.3, 8.4, 8.5).

8.4.2 Analysis of Specific Responses from the NIH

Figure 8.3 provides additional visibility into the views concerning Blastocystis pathogenicity. This correspondence contains several notable elements:

1. Fig 8.3 *"They have 2 nonpathogens…they should consider using metronidazole … or Trimethoprin/sulfa"*: The author suggests that both *Blastocystis* and *Dientamoeba fragilis* are nonpathogenic, but antimicrobial treatment is suggested. This statement may illustrate the complexity of defining the term "pathogenic."

Hall, Lee (NIH/NIAID) [E]

From: Hall, Lee (NIH/NIAID) [E]
Sent: Tuesday, June 08, 2010 9:59 AM
To: Giovanni, Maria (NIH/NIAID) [E]
Subject: RE: Blastocystis - biomedical relevance

Maria,
This is not a high priority.

Although *Blastocystis spp.* have a wide geographic distribution, it is not clear that they are in fact "true" pathogens - perhaps just a low grade pathogen, or just part of the normal gut flora. Clinical symptoms attributed to *B. hominis* infection include the usual panoply of GI complaints, e.g., diarrhea, bloating, flatulence, cramping. The evidence that *B. hominis* is responsible for disease is based largely on case reports and uncontrolled or retrospective studies of patients with non-specific GI symptoms. Neither the identification of *B. hominis* in stool or the number of organisms in stool has been correlated with symptoms, however. Some observational studies and small clinical trials of antibiotic therapy have suggested improvement of symptoms with disappearance/reduction in *B. hominis*, but since the antibiotics themselves are broad spectrum (e.g., Bactrim, metronidazole, nitazoxanide), it is hard to attribute the effects specifically due to antibiotic activity on *B. hominis* and not on other intestinal microbes. In any event, blastocystosis is often considered to be self-limited, and many physicians opt not to treat *B. hominis* when it is found in the stool even when signs/symptoms are present unless there is some underlying condition, e.g., immunocompromised hosts.

Happy to discuss further if you want.

Lee

From: Giovanni, Maria (NIH/NIAID) [E]
Sent: Tuesday, June 08, 2010 8:04 AM
To: Hall, Lee (NIH/NIAID) [E]
Subject: FW: Blastocystis - biomedical relevance

I may have already asked you this but is this something of importance to pursue. He is very persistent.

Maria Y. Giovanni, Ph.D.
Assistant Director for Microbial Genomics & Advanced Tech.
Division of Microbiology and Infectious Diseases
NIAID/NIH/DHHS
6610 Rockledge Dr. MSC 6603; Room 4807
Bethesda, MD 20892-6603
301-496-1884 Phone
301-480-4528 FAX
mgiovanni@niaid.nih.gov

Disclaimer:
The information in this e-mail and any of its attachments is confidential and may contain sensitive information. It should not be used by anyone who is not the original intended recipient. If you have received this e-mail in error please inform the sender and delete it from your mailbox or any other storage devices. National Institute of Allergy and Infectious Diseases shall not accept liability for any statements made that are sender's own and not expressly made on behalf of the NIAID by one of its representatives.

Fig. 8.5 An internal NIH communication discussing a researcher's inquiry as to whether the NIH would fund a proposed *Blastocystis* research project. The position of the NIH staff member is consistent with communications made to BRF by patients and researchers, which indicate that agency believes no evidence exists that the infection could cause illness in healthy humans

Mainly, some definitions of the term would suggest that if an antimicrobial treatment is prescribed for a microbial infection with the intention of relieving patient symptoms, then the microbe in question would be considered a pathogen.

2. Fig 8.3 *Blastocystis is not a parasite and has been reclassified as a fungus*: We were unable to find any reference to *Blastocystis* being reclassified from a parasite to a fungus in recent reviews of *Blastocystis*, or in common archives of medical literature. All recent papers continue to refer to *Blastocystis* as a parasitic infection, and none note any status as a fungus (Boorom et al. 2008;

Table 8.3 Institutional Affiliation of Selected Researchers co-authoring papers on *Blastocystis*

Agency	Author (year of publication)	Title
US National Institute of Health	Zierdt (1991a)	*Blastocystis Past and Future*
US Center for Disease Control	CDC (2000)	*Blastocystis Fact Sheet, 2000–2010*
US Center for Disease Control	CDC (2010)	*Blastocystis Fact Sheet, 2010-present*
US Center for Disease Control	Boorom et al. (2008)	*Oh my aching gut: irritable bowel syndrome, Blastocystis, and asymptomatic infection*
China Center for Disease Control	Boorom et al. (2008)	
US Air Force	Jones et al. (2009)	*Association of Blastocystis subtype 3 and 1 with patients from an Oregon community presenting with chronic gastrointestinal illness*
US Department of Agriculture	Santin et al. (2011)	Development of a new PCR protocol to detect and subtype *Blastocystis spp.* from humans and animals
World Health Organization (Coordinating Center)	Parkar et al. (2010)	Molecular characterization of *Blastocystis* isolates from zoo animals and their animal-keepers

Papers on the infection are commonly published by researchers at national and regional public health organizations and hospitals, as well as national universities. Within the USA, the NIH elected to discontinue research participation after the mid-1990s, but scientists from other federal agencies have published papers on the subject

Stensvold et al. 2010; Tan et al. 2011). The statement illustrates the difficulty associated with developing consensus around technical aspects of *Blastocystis* infection. Variations in viewpoints among individuals may develop spontaneously, and do not appear to originate from any medical studies.

3. Fig 8.3 *D. fragilis is nonpathogenic*: The statement was intriguing, as most recent papers from national health services in Canada, the UK, and large hospitals in Australia identify *D. fragilis* as pathogenic, and suggest that symptoms would be largely similar to those seen in other intestinal parasitic diseases (Lagace-Wiens et al. 2006; Windsor 2007; Windsor and Macfarlane 2005). Reviews noting diagnosis and treatment of patients are commonly published in mainstream clinical journals, such as the American Journal of Tropical Medicine and Hygiene (Stark et al. 2012) and Clinical Microbiology Reviews (Johnson et al. 2004).

Figures 8.4 and 8.5 illustrate an inquiry to the NIH concerning funding of a *Blastocystis* research grant. In Fig. 8.4, the researcher notes the trend of *Blastocystis* research, and the development of a consensus of researchers that the microbe is pathogenic. In Fig. 8.5, the discussion at the NIH concerning this grant suggests that no funds would be made available for a project investigating *Blastocystis* infection (Table 8.4).

Table 8.4 *Blastocystis* Firsts, 1990–2012

Item	Author (year)	Country
First studies to examine antimicrobial resistance	Silard and Burghelea (1985)	Romania
	Zaman and Zaki (1996)	Pakistan
First study to recommend that *Blastocystis* patients be diagnosed with IBS	Markell and Udkow (1986)	USA
First study to suggest that *Blastocystis* infection was the underlying cause of irritable bowel syndrome	Hussain et al. (1997)	Pakistan
First mouse model	Moe et al. (1997)	Singapore
First studies to suggest some *Blastocystis* isolates are associated with acute infection, while others may be associated with chronic infection	Lanuza et al. (1999)	Italy
	Stensvold et al. (2011)	Denmark
First study to report that common clinical diagnostic techniques fail to identify most *Blastocystis* infections	Leelayoova et al. (2002)	Thailand
First detailed study of a *Blastocystis* epidemic	Nimri (1993)	Jordan
First long term study (2 years) of *Blastocystis* immune response in patients with and without gastrointestinal symptoms	Kaneda et al. (2000)	Japan
First study to report a difference in *Blastocystis* antigen immune response between symptomatic and asymptomatic hosts	Mahmoud and Saleh (2003)	Egypt
First placebo controlled trial for treatment of *Blastocystis* infection with antimicrobials	Nigro et al. (2003)	Italy
First study to report death in animal models following experimental infection of *Blastocystis* from symptomatic humans	Yao et al. (2005)	China
First major analysis of *Blastocystis* genotypes from multiple countries	Noel et al. (2005)	France
First large epidemiological study of *Blastocystis* infection	Amin (2005)	USA
First large scale population study of *Blastocystis* using a reliable diagnostic technique and sub-typing of isolates	Li et al. (2007a)	China/Japan
First comparison of conventional diagnostics, stool culture and PCR detection for identification of *Blastocystis* infection in stool samples	Stensvold et al. (2007a)	Denmark
First international agreement on naming conventions for *Blastocystis* isolates	Stensvold et al. (2007b)	Denmark
First studies to investigate waterborne transmission of *Blastocystis*, including molecular epidemiology	Li et al. (2007b)	China
	Leelayoova et al. (2008)	Thailand
	Eroglu and Koltas (2010)	Turkey
First real-time PCR test for *Blastocystis* infection	Jones Ii et al. (2008)	USA
First genotyping of *Blastocystis* infection in cancer patients	Tan et al. (2009)	Malaysia
First study to provide details (molecular weights of antigens) for *Blastocystis* immune response in symptomatic humans	Hegazy et al. (2008)	Egypt

(continued)

Table 8.4 (continued)

Item	Author (year)	Country
First study to perform experimental animal infection with multiple *Blastocystis* isolates that were sub-typed	Hussein et al. (2008)	Egypt
First systematic review of *Blastocystis* literature	Boorom et al. (2008)	USA + ten other countries
First detailed clinical reports of *Blastocystis* in producing skin rash	Katsarou-Katsari et al. (2008)	Greece Turkey
First sub-typing of *Blastocystis* infection in an irritable bowel syndrome (IBS) patient as well as a "Gulf War Illness" patient	Jones et al. (2009)	USA
First study to genotype *Blastocystis* in inflammatory bowel disease patient	Dogruman-Al et al. (2010)	Turkey
First study to report multiple immunosuppressive properties of *Blastocystis* secretory substances	Chandramathi et al. (2010a)	Malaysia
First study to report that *Blastocystis* (and E. histolytica) are associated with relapsing symptoms in inflammatory bowel disease patients	Yamamoto-Furusho and Torijano-Carrera (2010)	Mexico
First study to identify a diagnostic test whose results differ in both symptomatic *Blastocystis* patients, and in experimentally infected rats	Chandramathi et al. (2009, 2010b)	Malaysia
First study to report that *Blastocystis* down-regulated nitric oxide production	Mirza et al. (2011)	Singapore
First complete genome sequencing of *Blastocystis*	Denoeud et al. (2011)	France
First study to report that DNA extraction kits have greatly varying performance level in their ability to produce DNA for *Blastocystis* detection	Yoshikawa et al. (2011)	Turkey/Japan
First study to identify a host genetic trait which is responsible for mediating symptoms in *Blastocystis* infection	Olivo-Diaz et al. (2012)	Mexico

Participation of laboratories in Asia, the Middle East, and most recently Europe and Mexico has been essential for the development of *Blastocystis* research. The lack of US participation after 1995 is notable

8.5 Blastocystis and IBS in Medical Studies

Published medical studies provide another avenue for understanding how views about pathogenicity can vary between individuals, and how those views are formed. The two studies by Dr. Edward Markell, provide a good reference point, as Dr. Markell's 1986 paper, "Blastocystis: Pathogen or Fellow Traveler" remains one of the ten most cited *Blastocystis* studies. The abstract of the paper describes his conclusions as follows:

> To investigate this possibility, we identified 148 persons whose stools contained this organism. Of this number, 32 had at least 6 stool examinations performed. Twenty-seven of the 32 persons were later found to have greater than or equal to 1 recognized pathogens—*Entamoeba histolytica, Giardia lamblia* or *D. fragilis*—and, after receiving appropriate therapy,

> became asymptomatic. The *B hominis* infection, however, was unaffected by therapy. Five persons with only *B. hominis* infection were treated with iodoquinol without effect; these persons fulfilled the medical criteria for irritable bowel syndrome. We believe that when an apparently symptomatic *B. hominis* infection responds to therapy, the improvement probably represents elimination of some other undetected organism causing the infection.
> *Blastocystis: Pathogen or Fellow-Traveler* (Markell and Udkow 1986)

We can identify several concepts about pathogenicity from this statement:

1. *Pathogenicity can be determined by response to antimicrobial treatment.* That is, the lack of symptomatic improvement following treatment with iodoquinol shows that a microbe is nonpathogenic. Subsequent in vitro studies showed the iodoquinol had little activity against *Blastocystis* infection (Dunn and Boreham 1991; Mirza et al. 2010). The idea that pathogens must be treatable is also noted in a 1990 letter to the Journal of Clinical Microbiology concerning *Blastocystis* pathogenicity (Rosenblatt 1990), and also in a 2010 review on the treatment prescribed for *Blastocystis*, with the title, "Eradication of Blastocystis carriage with antimicrobials: reality or delusion" (Stensvold et al. 2010).
2. *If a patient has symptoms meeting the criteria for irritable bowel syndrome, microbial causes can be eliminated.*
3. *The drop-out effect is not a concern in studies of this nature.* Most patients (116/148, 78 %) dropped out of Dr. Markell's study. Of the original 148 study participants, only 32 completed the study. In the paper text, Markell noted that he communicated the results of the testing with patients while the study was being conducted, causing patients with *Blastocystis* mono-infections to drop out: "In others [leaving the study], the patient was reassured by initial examinations which were negative for recognized pathogens, and did not return." (Markell and Udkow, 1986, p. 1024) Markell did not perceive this drop-out effect to be incompatible with the study's conclusion that symptomatic *Blastocystis* mono-infections were uncommon.
4. *The appearance of another pathogen during the study proves that the original infection was not the cause of illness in the patient.* Markell collected six samples over a period of 11 months, according to the paper.
5. *At the end of Markell's study, he was left with a population where 15.6 % (5/32) of the Blastocystis patients were symptomatically mono-infected.* This level of symptomatic mono-infection is thought to prove that *Blastocystis* does not cause of illness.

The issue as to whether asymptomatic carriers can be used to prove that a microbial infection is nonpathogenic repeatedly appears in the literature concerning *Blastocystis* infection. The argument is that since all enteric protozoal infections have high rates of asymptomatic carriers, this would not be the best criteria to use in "proving" that a suspected enteric protozoan is nonpathogenic. Dr. Charles H. Zierdt's response to the issue of asymptomatic carriers was as follows:

> Dr. Rosenblatt states: "Some patients with diarrhea have the organism in their stool and some do not; some asymptomatic patients have it in their stool and some do not." What a beautiful description of the presence of *Entamoeba histolytica*!
> *Blastocystis pathogenicity* (Zierdt 1991b)

The issue as to whether *Blastocystis* patients (and patients with other microbial infections) should be diagnosed with IBS continues to be controversial in the medical community. One group of scientists indicates that such infections should be excluded before diagnosing patients with IBS, publishing a 2007 review article entitled, "*Irritable bowel syndrome: a review on the role of intestinal protozoa and the importance of their detection and diagnosis*" (Stark et al. 2007). However, other researchers argue the microbial causes for gastrointestinal illness have been given too much emphasis in the medical community. Dispensing an infective diagnosis for diarrhea based solely on the presence of an infection, be it bacterial or protozoal, is not appropriate because infections appear in asymptomatic patients:

> For example, an individual with a bacterial gastroenteritis or other bowel disorder who has no concurrent psychosocial difficulties and good coping skills may not develop the clinical syndrome (or be aware of it) or if it does develop, may not perceive the need to seek medical care. Another individual with coexistent psychosocial comorbidities, high life stress, abuse history, or maladaptive coping, may develop a syndrome (e.g., postinfectious irritable bowel syndrome [IBS] or dyspepsia), go to the physician frequently, and have a generally poorer outcome."
>
> *The Functional Gastrointestinal Disorders and the Rome III Process* (Drossman 2006)

This text appears in the introduction to the Rome III Process document, published in the journal *Gastroenterology*. The document is an international standard for the diagnosis of functional gastrointestinal disorders. According to the text, the paradigm shift began thirty years ago, and was produced as a result of centuries-long unjust exclusion of a type of research by the medical community, a process which is remedied with the development of new paradigms:

> The first event began 3 decades ago with a paradigm shift that moved away from conceptualizing illness and disease based on a 3-century-old reductionistic model of disease in which the effort was to identify a single underlying biological etiology to a more integrated, biopsychosocial model of illness and disease. The former disease-based model had its roots with Descartes' separation of mind and body and at the time was a concept that harmonized prevailing societal views of separation of church and state. What resulted was permission to dissect the human body (which was previously forbidden), so disease was defined by what was seen (i.e., pathology based on abnormal morphology). This approach led to centuries of valuable research producing effective treatments for many diseases. The concept of the mind (i.e., the central nervous system [CNS]) as being amenable to scientific study or as playing a role in illness and disease was marginalized, however. The mind was considered the seat of the soul, not to be tampered with.
>
> *The Functional Gastrointestinal Disorders and the Rome III Process* (Drossman 2006)

The viewpoint that individuals control their response to infectious diseases is not a new idea, and was promoted widely in the teachings of Mary Baker Eddy, the founder of the Christian Science movement in the USA, which advocates prayer for the treatment of infectious diseases (Swan 1983). Advocates of the idea point to the presence of asymptomatic carriers of infectious diseases like *Vibrio cholerae*. The paradigm can be extended to other diseases, such as tuberculosis. The following appeared in a journal of the largest charity in the UK devoted to patients with irritable bowel syndrome (IBS):

> We have known since the birth of civilization how people weakened by malnutrition and poverty are particularly susceptible to illness. The same applies to people undermined by

emotional distress. The dramatic potential of the Tubercle bacillus to devastate the health of young men and women crossed in love is a dramatic theme of many novels of the nineteenth and early twentieth centuries.

Nick Reed, MD, Issue 69, *Gut Reaction, Journal of the Gut Trust*

It is difficult to resolve this controversy between microbiologists and bio-psycho-social (BPS) scientists because of the different methods used by each group. Specifically, BPS faults the use of reductionism in the medical community, a process which typically involves narrowing possible causes with laboratory experiment, until the smallest set of factors is identified which can reproduce the disease. BPS scientists emphasize "multi-determinism," and believe many factors influencing symptoms in infectious diseases cannot be clearly quantified, and have unfairly been excluded from the medical process. These influences might include traumatic experiences early in life, maternal separation, the cumulative effect of emotionally stressful life events, etc. BPS scientists may also draw on discussion of novels or seventeenth century European philosophers, while microbiologists emphasize repeatable clinical and laboratory studies.

8.6 Prevalence as a Test for Pathogenicity

Measurements of the prevalence of *Blastocystis* have also been cited in conjunction with discussion of pathogenicity. In 2006, I worked with an Oregon State legislator, Representative Sarah Gelser, to bring a bill, HB2699, before the legislature to make *Blastocystis* a reportable infection (Fig. 8.6).[3] The bill was opposed by the State Epidemiologists who cited the prevalence of the infection as proof that it could not cause disease:

Based on the number of studies we have seen or read, about 10–20 % of the population in the world, and the USA, may actually carry Blastocystis hominis. That would equal about 700,000 individuals in Oregon carrying the organism, walking around with Blastocystis hominis, and not having any symptoms at all. So that is very telling of the likelihood of the organism and the possibility of causing disease.

Dr. Emilo deBess, Testimony to State of Oregon Health Policy Committee, March 2007

A counter-point to the argument that prevalent microbes can not be pathogenic was suggested in a 2010 review on *Blastocystis*, which noted that the prevalence of chronic gastrointestinal illness (IBS) generally tracked the prevalence of

[3] The logic was that physicians already recognized *Blastocystis* as a pathogen by their actions. The common practice, as communicated to me by interviews with multiple physicians and gastroenterologists, was to treat *Blastocystis* when it was found in symptomatic patients. That is, laboratories looked for the organism in stool samples, physicians diagnosed patients with blastocystosis, and treated the infection. We had detailed testimony from patients or their guardians who had been seen repeatedly by physicians associated with the major hospitals and clinics in the region, and had been diagnosed and treated for *Blastocystis* infection. All physicians and board certified gastroenterologists who we contacted in the area supported passage of the bill, and provided signed letters supporting it.

Fig. 8.6 The author testifying to the State of Oregon Health Policy Committee in support of making *Blastocystis* a reportable infection in Oregon

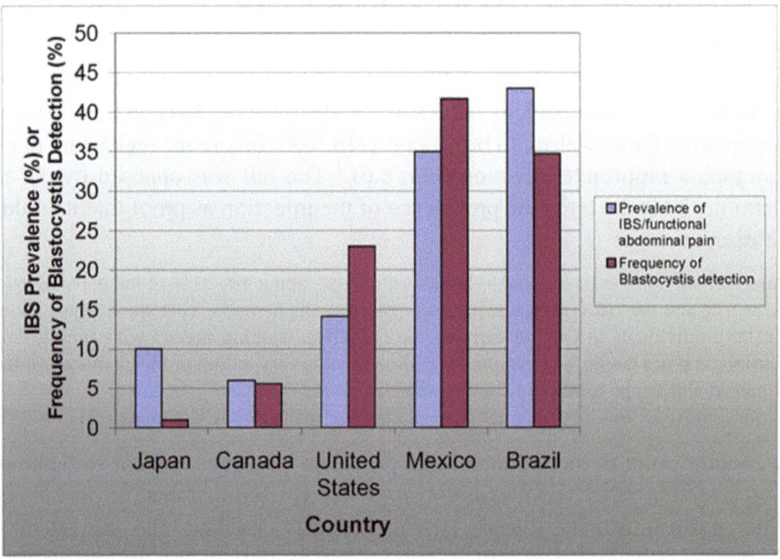

Fig. 8.7 Comparison of the prevalence of irritable bowel syndrome (IBS) or chronic abdominal pain in surveys from various countries to the frequency of detection of *Blastocystis* infection at clinical laboratories. The high prevalence of both the patients and the illness may contribute to opposition to pathogenicity, some individuals suggest that an infection which is highly prevalent is unlikely to be pathogenic

Blastocystis infection in a number of developed nations (Fig. 8.7) (Boorom et al. 2008). The review also noted that in Egypt, 21 % of apparently individuals have been found to carry *E. histolytica*. In a study of a slum in Brazil, 11 % carried *E.*

histolytica. In both cases, ELISA assays were used to differentiate *E. histolytica* from *E. dispar*, and the number reported is the prevalence of *E. histolytica*.

Measurement of the prevalence of the infection in specific populations has also been cited as a way to determine pathogenicity of *Blastocystis*. Markell and Udkow's second study, published in 1993, noted that the infection was found at a similar prevalence in symptomatic and asymptomatic individuals:

> There was no statistically significant difference in prevalence between the two study groups (11.5 % vs. 12.8 % for asymptomatic and symptomatic respectively, $P=0.435$, Fisher's one-tailed test)
>
> *Blastocystis hominis: Prevalence in Asymptomatic versus Symptomatic Hosts* (Udkow and Markell 1993)

However, it has been noted that Markell also published two papers noting that there was no correlation between the presence of *E. histolytica* and *G. intestinalis* and symptoms in his study groups (Boorom et al. 2008).

> Of the health fair group, however, 44 (42 %) responded "no" to all nine questions [concerning gastrointestinal symptoms]. Of the 44, 12 (27 %) were infected with potential pathogens. No correlation between symptoms and the presence or absence of infection with *E. histolytica*, *G. intestinalis* could be detected.
>
> *Intestinal parasitic infections in homosexual men at a San Francisco Health fair*, (Markell et al. 1983)
>
> The prevalence of Entamoeba histolytica (28.6 %) was similar to that reported in other studies, whereas that of Giardia lamblia was lower. Infection with E. histolytica was correlated significantly with a prior history of syphilis or gonorrhea ($P < 0.0001$), with the number of sexual partners in the preceding 12-month period ($P < 0.0001$), and with the reported frequency of oral–anal sexual contact ($P < 0.001$). Giardial infection was also significantly related to oral–anal sex ($P < 0.001$). No relation was seen between the presence or absence of gastrointestinal symptoms and infection with pathogenic protozoa.
>
> *Intestinal protozoa in homosexual men of the San Francisco Bay area: prevalence and correlates of infection* (Markell et al. 1984)

Markell's last study equated attention to *Blastocystis* with dereliction of the physician's professional responsibility:

> Reconciling even potential pathogenicity with the essentially equal prevalence of this organism in symptomatic and asymptomatic persons...seems difficult. Furthermore, focusing attention on *B. hominis* may stifle further investigation to the possible detriment of the patient or may prompt treatment that is both unnecessary and potentially dangerous.
>
> Blastocystis hominis: Prevalence in Asymptomatic vs. Symptomatic Hosts (Udkow and Markell 1993)

We do not have an explanation as to why investigating *Blastocystis* would cause such harm, while investigating *G. intestinalis* and *E. histolytica* would presumably be acceptable, even though as Markell noted, there was similarly no correlation between any of these infections and symptoms in his study populations. That is, we have an example where the identical finding is interpreted in different ways by the same researcher, depending on the objectives of the paper.

Why was every intestinal protozoal infection uncorrelated with symptoms in Markell's studies? Studies of intestinal protozoal infections in humans are, at best, partially controlled studies. Both Markell and a *Blastocystis* review paper suggested the possibility that patients who show symptoms are more likely to seek medical treatment, which can skew results of population studies (Boorom et al. 2008; Markell et al. 1983). That is, if all the symptomatic patients seek and receive treatment, we are left with asymptomatic patients for epidemiologists to study. Besides patient behavior, acquired immunity can influence expression of symptoms. Studies citing a lack of correlation between intestinal protozoal infections and symptoms are often conducted on individuals with frequent exposure to the infections (Pickering et al. 1984). The results from detailed study of individuals in endemic settings suggests the possibility that in cases of frequent exposure, most of the infections are asymptomatic, and this process may maintain immunity at a level such that the infection does not develop into severe illness (Haque et al. 2006). This pattern is very different from some other diseases (i.e., HIV infection, tuberculosis) where individuals who are most frequently exposed also have the highest morbidity.

Other confounding factors include the presence of an undetected pathogen like *Cryptosporidium* spp. in the microbe-free symptomatic population (Chapter 7, Section 7.13); the presence of innate immunity (Duggal et al. 2004; Haque et al. 2002); and the manner in which the physician classifies a patient as symptomatic or asymptomatic (Janoff et al. 1990). Markell was apparently aware of the influence of acquired and innate immunity in mediating symptoms of parasitic infections, as he noted both in his text on medical parasitology, but they were not noted as possible factors in influencing the outcome of *Blastocystis* population studies (Markell and Voge 1976).

8.7 Other Tests for Pathogenicity

Additional tests for pathogenicity have been proposed. One author, writing a letter to the editor in the Journal of Clinical Microbiology in 1990, cited five criteria:

> Markell and Udkow comment on the "guilt by association" phenomenon which has grown up around this organism in the recent literature. Stated another way, simply repeating continuously the statement that *B. hominis* is a pathogen will not make it so. In addition to the fact that there is no clear segregation of this organism between symptomatic and asymptomatic persons, we need to remember the following: (1) Koch's postulates have never been satisfied (there is no reproducible model of experimental infection due to *B. hominis* only), (2) no pathologic evidence of or immunologic response to "infection" has been demonstrated in humans, (3) no mechanisms of pathogenicity, such as toxin elaboration, attachment to intestinal mucosa, or invasiveness, in humans have been described, (4) no antimicrobial agent which is uniquely active against *B. hominis* has been shown to reliably eradicate both the organism and the diarrhea, and (5) there has never been a point-source outbreak of gastroenteritis in which epidemiologic evidence suggested that *B. hominis* was the cause.
>
> *Blastocystis Pathogenicity* (Rosenblatt 1990)

In this case, the author suggests that a combination of laboratory studies (animal models, immunologic response, and pathogenesis) should be used in conjunction with an epidemiological criterion of the point-source outbreak, and the clinical criteria of a drug that acts against the microbe.

Additional criteria were cited by Dr. Emilio deBess, in testimony from the State of Oregon Epidemiologist's Office in testimony concerning legislation making *Blastocystis* a reportable infection (Supplemental files).

1. We asked the CDC. They have an epidemiologist section where they deal with things like *Giardia* and other parasites, and again they claim there is no convincing evidence at this point that would make us think that *B. hominis* is an organism that could actually cause an illness or disease.
2. One of the interesting facts about this particular organism or parasite is that when it was actually fed to animals that were bacteria free, it did not cause disease that is one of the basis by which we actually look at how a bacteria or parasite works.
3. As I said, the AIDS and HIV era did not put Blastocystis hominis forward as an organism that could actually cause illness in people who are immune compromised.
4. There are few lines of evidence in regards to blastocystosis but one of them in 2000 has to do with a Swedish study, and they actually found more Blastocystis hominis in well people than sick people. And that is kind of a landmark article.

– Dr. Emilio deBess, DVM
Oregon State Epidemiologist's Office
April 2006

As an illustration of the difficulty in developing consensus on the idea of pathogenicity, Dr. deBess' four criteria form almost a disjoint set with respect to Rosenblatt's criteria. Dr. deBess introduces a new criterion for establishing a pathogen—mainly, it should be shown to be problematic in HIV patients. Also, while Dr. Rosenblatt limited his criteria to the existence of medical studies, Dr. deBess introduces a criterion for opinion polling, in that a specific group should agree that the organism is pathogenic. While Rosenblatt's criteria required that specific studies exist, Dr. deBess criteria include what might be called "lack of existence" criteria. That is, an organism should not be considered pathogenic if specific studies exist which would conflict with the idea of pathogenicity, such as the Swedish study, or the animal study.

The abstract from an invited review on *Blastocystis* published in the January 2012 issue of *Clinical Infectious Diseases*, the journal of the Infectious Disease Society of America, provides some additional viewpoints on pathogenicity, and also shows how the requirements can potentially change over time:

Parasites in the genus *Blastocystis* comprise several subtypes (genotypes) and have a worldwide distribution. In some surveys, these are the most common parasites found in human stool specimens. An emerging literature suggests that the pathogenicity of *Blastocystis* is related to specific subtypes and parasite burden, although even individuals with small numbers of cysts may be symptomatic. Some data suggest an association between infection with *Blastocystis* and irritable bowel syndrome. However, there anuare few clinical studies demonstrating a direct relationship between the presence of this parasite and disease, few animal models to explore this relationship, and no consensus as to appropriate treatment. We recommend that asymptomatic individuals with few cysts not be treated. However, those who have gastrointestinal or dermatologic signs and symptoms

> and many cysts in stool specimens may require treatment. Metronidazole is the drug of choice. Additional studies are required to determine pathogenicity and appropriate therapy.
> *Blastocystis: To Treat or Not To Treat* (Coyle et al. 2012)

In 2006, animal models had been identified only by a research team in Singapore (Moe et al. 1997). By 2012, a number of laboratories, primarily in the Middle East and China had published papers describing animal models for blastocystosis in mice and rats. While the body of research would have met Rosenblatt's criteria from 1990 for animal models, a new criterion is proposed by Coyle, mainly that many animal models should exist (Table 8.5).

At the same time, some researchers had reported that symptomatic experimental animal infection required inoculating animals orally with a large number of cysts, noting that small numbers of cysts did not produce symptomatic infection (Elwakil and Hewedi 2010). This adds a level of complexity if negative criteria are to be applied, such as that noted by Dr. deBess in 2006. Mainly, if researchers use a small number of cysts in inoculating animals, Dr. deBess' second criteria (or objection) will be present, mainly that studies exist which show that inoculating animals with the microbe does not produce illness.

Table 8.6 shows the diversity of opinion concerning pathogenicity that can exist among experts by comparing nine different criteria for pathogenicity cited by these five experts (Markell, Rosenblatt, deBess, Hall, Coyle). At most, different experts share only two of these nine properties, and in many cases, experts do not have any concepts in common. Pathogenicity appears to be highly subjective, and the discovery of new information does not necessarily improve the level of consensus between decision makers. This may explain why historically, decision making about many pathogens is seldom performed by organizations of physicians.

8.8 Other Methods for Making Medical Decisions

The slow pace of progress in the area of gastrointestinal protozoal infections may be due to the mechanism by which public policy decisions are made. Although medical studies go through a peer review process prior to publication, in many cases, the development of policy from medical studies is performed through a second review process (Table 8.7). Examples include the process of drug approval, which is performed by the Food and Drug Administration (FDA) in the USA. The process by which mental disorders are defined is performed by panels of experts convened every few years by the American Psychiatric Association.

In these cases, it is recognized that the second review process is needed, because a consensus does not develop in the community solely from the existence of more studies. The process of reviewing studies and making decisions requires resources and an environment which may not be present in the physician's office. It is not a process that happens spontaneously. The process is designed to prevent specific biases. For example, in drug approval, physicians who develop drugs may develop biases in favor of their use, so the FDA selects groups of individuals to review studies

Table 8.5 Comparison of criteria for pathogenicity from different experts, based on statements from those individuals. Between the five experts examined, at least nine different criteria for pathogenicity are suggested, but very few experts identify the same criteria for pathogenicity

Authority	Year	Prevalence in symptomatic and asymptomatic	Existence of animal models	Severe illness in AIDS patients	Success of iodoquinol treatment	Success of any treatment	Physician behavior (do they treat it?)	Number of animal models	Presence of epidemic	Must be present at low rate in general population
Markell	1986	Yes	No	No	Yes	Yes	No	No	No	No
Rosenblatt	1990	No	Yes	No	No	Yes	No	No	Yes	No
deBess	2009	Yes	Yes	Yes	No	Yes	Yes	No	No	Yes
Hall	2010	No	No	No	No	No	Yes	No	No	No
Coyle	2012	No	Yes	No	No	Yes	No	Yes	No	No

Table 8.6 Number of times overlapping is observed in criteria for pathogenicity cited by experts, based on data from Table 8.5. Out of nine possible criteria only two criteria were identified by more than one researcher, suggesting that the idea of pathogenicity may be a strongly subjective concept.

	Markell	Rosenblatt	deBess	Hall	Coyle
Markell	–	1	2	0	1
Rosenblatt	–	–	2	0	2
deBess	–	–	–	1	2
Hall	–	–	–	–	0
Coyle	–	–	–	–	–

Table 8.7 Examples of groups that review medical studies regularly, and public policy decisions concerning those studies

Region	Group name	How often	Input	Output
USA	Food and drug administration (FDA)	Weekly–monthly	Medical studies testimony	Decisions concerning drug approval and warning labels
USA	Institute of medicine (IOM)	Several times/year	Medical studies testimony	Written medical policy analyses, often for US Congress
USA	American Psychiatric association (APA)	Once every several years	Medical studies	Statistical manual of mental disorders (DSM)—mental disorders and diagnosis codes
Europe	European medicines agency	Weekly–monthly	Medical studies	Decisions concerning drug approval and warning labels for the EU

who are not involved in the drug's development. In writing the Statistical Manual of Mental Disorders (DSM), the American Psychiatric Association recognizes that different physicians may develop different ideas about how to classify mental disorders, and that there is a value in developing an industry-wide standard.

Why are such regulatory organizations necessary in medical science, and not in physics or mathematics? One reason may come from an examination of the number of individuals who have an interest in the outcome of a decision, and the process by which the "debate" is held. In the case of *Blastocystis* infection in the USA, there have been approximately five researchers who have pursued multiyear research projects concerning *Blastocystis* infection over the last 30 years. However, data from those research efforts impacts medical practice of virtually all physicians in the USA, who number over 600,000.

The decision making process is generally self-selective, with researchers choosing to publish, and individuals within the medical community choosing to object to such publication. If just one physician in 1000 objects to a type of research, individuals who object to the research will quickly out-number the parasitologists. In contrast, organizations that develop medical policies generally use an appointment process, where individuals are selected to serve on a committee which renders a decision.

The mechanics of this process might be seen in the recent controversy surrounding reclassification of Pluto from a planet to a small planetary body. Even though this proposal would have little tangible economic or social impact, the reclassification was vigorously opposed by many schoolchildren and teachers. State legislative bodies in the USA even passed laws declaring Pluto to be a planet. The final decision was made by the International Astronomical Union, and is a potential illustration of the necessity of decision making bodies in science. It is unlikely that publication of additional studies about the size of Pluto would have much effect in resolving the controversy.

8.9 Blastocystis Firsts and the Shift of Research East

In compiling a list of significant events in *Blastocystis* research (Table 8.4), it is intriguing to note the number of countries involved, and that regions in the Middle East and Asia are strongly represented in this list, while the USA and UK are virtually absent. In the development of an international standard for naming *Blastocystis* isolates in 2010, no US researcher participated.

It is suggested that this distribution differs from that seen in *G. intestinalis* or *Cryptosporidium spp.*, where US and UK teams made significant contributions. Researchers from the US and UK were the first to achieve a variety of milestones, such as the first animal models (Fantham and Porter 1916), the first studies of immunological responses in humans (Brown et al. 1973), leadership for the first full-genome mapping of *G. intestinalis* (Morrison et al. 2007) and *Cryptosporidium* spp.(Xu et al. 2004). In the case of *Blastocystis*, the US withdrawal from research has altered the landscape for this organism, and most advanced research is now performed in Asia, the Middle East, and more recently Europe and Mexico (Table 8.4).

The variation cannot be explained by the prevalence of the infection alone, since studies show that *Blastocystis* ranks as the most frequently identified parasitic infection in the US and UK, and is detected in over 20 % of individuals with gastrointestinal illness in studies from those countries (Windsor et al. 2007, Boorom et al. 2008). Given research trends noted earlier, some parallels may be drawn between the situation in *Blastocystis*, and the ban of US federal funding of many types of embryonic stem cell research, which was a response to objections from leaders of certain religious groups in the USA. The ban provides advantages to research centers in Asian countries, where similar objections do not exist (Walters 2004).

For researchers interested in future work, this suggests that the epicenter for this type of research will remain in the East, as the majority of advanced research will come from a few European countries (Denmark, Turkey, France, Greece) working with Asian and Middle Eastern groups. Those collaborations can already be seen in papers, such as a 2010 review paper with strong participation from researchers in these regions (Boorom et al. 2008), a collaboration between French and Egyptian researchers (Souppart et al. 2010), and a collaboration between Turkish and Japanese researchers (Yoshikawa et al. 2011).

References

Amin O (2005) Epidemiology of *Blastocystis hominis* in the United States. Research Journal of Parasitology 1(1):1–10, 2006

Amin OM (2002) Seasonal prevalence of intestinal parasites in the United States during 2000. Am J Trop Med Hyg 66(6):799–803

Bazerman MH, Moore DA (2009) Judgment in managerial decision making, 7th edn. John Wiley & Sons, Hoboken, NJ

Boorom KF, Smith H, Nimri L, Viscogliosi E, Spanakos G, Parkar U, Li LH, Zhou XN, Ok UZ, Leelayoova S, Jones MS (2008) Oh my aching gut: irritable bowel syndrome, *Blastocystis*, and asymptomatic infection. Parasit Vectors 1(1):40

Brown WR, Lansford CL, Hornbrook M (1973) Serum immunoglobulin E (IGE) concentrations in patients with gastrointestinal disorders. Am J Dig Dis 18(8):641–645

Cartlidge E (2012) Loose cable may unravel faster-than-light result. Science 335(6072):1027

CDC (2000) US Center for Disease Control Fact Sheet on *Blastocystis* archived on ARCHIVE.ORG. In. http://web.archive.org/web/20100308092248/http://www.cdc.gov/ncidod/dpd/parasites/blastocystishominis/factsht_blastocystis_hominis.htm

CDC (2010) US Center for Disease Control Fact Sheet on *Blastocystis* (current). In. http://www.cdc.gov/parasites/blastocystis/

Chandramathi S, Suresh K, Anita ZB, Kuppusamy UR (2009) Elevated levels of urinary hydrogen peroxide, advanced oxidative protein product (AOPP) and malondialdehyde in humans infected with intestinal parasites. Parasitology 136(3):359–363

Chandramathi S, Suresh K, Kuppusamy UR (2010a) Solubilized antigen of *Blastocystis hominis* facilitates the growth of human colorectal cancer cells, HCT116. Parasitol Res 106(4):941–945

Chandramathi S, Suresh K, Shuba S, Mahmood A, Kuppusamy UR (2010b) High levels of oxidative stress in rats infected with *Blastocystis hominis*. Parasitology 137(4):605–611

Coyle CM, Varughese J, Weiss LM, Tanowitz HB (2012) *Blastocystis*: to treat or not to treat. Clin Infect Dis 54(1):105–110

Crow MM (2011) Time to rethink the NIH. Nature 471(7340):569–571

Denoeud F, Roussel M, Noel B, Wawrzyniak I, Da Silva C, Diogon M, Viscogliosi E, Brochier-Armanet C, Couloux A, Poulain J, Segurans B, Anthouard V, Texier C, Blot N, Poirier P, Choo NG, Tan KS, Artiguenave F, Jaillon O, Aury JM, Delbac F, Wincker P, Vivares CP, El Alaoui H (2011) Genome sequence of the stramenopile *Blastocystis*, a human anaerobic parasite. Genome Biol 12(3):R29

Dogruman-Al F, Simsek Z, Boorom K, Ekici E, Sahin M, Tuncer C, Kustimur S, Altinbas A (2010) Comparison of methods for detection of *Blastocystis* infection in routinely submitted stool samples, and also in IBS/IBD Patients in Ankara, Turkey. PLoS One 5(11):e15484

Drossman DA (2006) The functional gastrointestinal disorders and the Rome III process. Gastroenterology 130(5):1377–1390

Duggal P, Haque R, Roy S, Mondal D, Sack RB, Farr BM, Beaty TH, Petri WA Jr (2004) Influence of human leukocyte antigen class II alleles on susceptibility to *Entamoeba histolytica* infection in Bangladeshi children. J Infect Dis 189(3):520–526

Dunn LA, Boreham PF (1991) The in-vitro activity of drugs against *Blastocystis hominis*. J Antimicrob Chemother 27(4):507–516

Eisenberg JN, Lei X, Hubbard AH, Brookhart MA, Colford JM Jr (2005) The role of disease transmission and conferred immunity in outbreaks: analysis of the 1993 *Cryptosporidium* outbreak in Milwaukee, Wisconsin. Am J Epidemiol 161(1):62–72

Elwakil HS, Hewedi IH (2010) Pathogenic potential of *Blastocystis hominis* in laboratory mice. Parasitol Res 107(3):685–689

Engelhardt HT, Caplan AL (1987) Scientific controversies: case studies in the resolution and closure of disputes in science and technology. Cambridge University Press, Cambridge, New York

Eroglu F, Koltas IS (2010) Evaluation of the transmission mode of *B. hominis* by using PCR method. Parasitol Res 107(4):841–845

Fantham HB, Porter A (1916) The pathogenicity of *Giardia (lamblia) intestinalis* to men and to experimental animals. Br Med J 2(2900):139–141

Ginther DK, Schaffer WT, Schnell J, Masimore B, Liu F, Haak LL, Kington R (2011) Race, ethnicity, and NIH research awards. Science 333(6045):1015–1019

Haque R, Duggal P, Ali IM, Hossain MB, Mondal D, Sack RB, Farr BM, Beaty TH, Petri WA Jr (2002) Innate and acquired resistance to amebiasis in bangladeshi children. J Infect Dis 186 (4):547–552

Haque R, Mondal D, Duggal P, Kabir M, Roy S, Farr BM, Sack RB, Petri WA Jr (2006) *Entamoeba histolytica* infection in children and protection from subsequent amebiasis. Infect Immun 74(2):904–909

Hegazy MM, Maklouf LM, El Hamshary EM, Dawoud HA, Eida AM (2008) Protein profile and morphometry of cultured human *Blastocystis hominis* from children with gastroenteritis and healthy ones. J Egypt Soc Parasitol 38(2):453–464

Hlavsa MC, Watson JC, Beach MJ (2005) Giardiasis surveillance–United States, 1998–2002. MMWR Surveill Summ 54(1):9–16

Huizenga JR (1992) Cold fusion: the scientific fiasco of the century. University of Rochester Press, Rochester, N.Y

Hussain R, Jaferi W, Zuberi S, Baqai R, Abrar N, Ahmed A, Zaman V (1997) Significantly increased IgG2 subclass antibody levels to *Blastocystis hominis* in patients with irritable bowel syndrome. Am J Trop Med Hyg 56(3):301–306

Hussein EM, Hussein AM, Eida MM, Atwa MM (2008) Pathophysiological variability of different genotypes of human *Blastocystis hominis* Egyptian isolates in experimentally infected rats. Parasitol Res 102(5):853–860

Janoff EN, Mead PS, Mead JR, Echeverria P, Bodhidatta L, Bhaibulaya M, Sterling CR, Taylor DN (1990) Endemic Cryptosporidium and *Giardia lamblia* infections in a Thai orphanage. Am J Trop Med Hyg 43(3):248–256

Johnson EH, Windsor JJ, Clark CG (2004) Emerging from obscurity: biological, clinical, and diagnostic aspects of *Dientamoeba fragilis.* Clin Microbiol Rev 17(3):553–570, table of contents

Jones Ii MS, Ganac RD, Hiser G, Hudson NR, Le A, Whipps CM (2008) Detection of *Blastocystis* from stool samples using real-time PCR. Parasitol Res 103(3):551–557

Jones MS, Whipps CM, Ganac RD, Hudson NR, Boorom K (2009) Association of *Blastocystis* subtype 3 and 1 with patients from an Oregon community presenting with chronic gastrointestinal illness. Parasitol Res 104(2):341–345

Kaneda Y, Horiki N, Cheng X, Tachibana H, Tsutsumi Y (2000) Serologic response to *Blastocystis hominis* infection in asymptomatic individuals. Tokai J Exp Clin Med 25(2):51–56

Kappus KD, Lundgren RG Jr, Juranek DD, Roberts JM, Spencer HC (1994) Intestinal parasitism in the United States: update on a continuing problem. Am J Trop Med Hyg 50(6):705–713

Katsarou-Katsari A, Vassalos CM, Tzanetou K, Spanakos G, Papadopoulou C, Vakalis N (2008) Acute urticaria associated with amoeboid forms of *Blastocystis* sp. subtype 3. Acta Derm Venereol 88(1):80–81

Lagace-Wiens PR, VanCaeseele PG, Koschik C (2006) *Dientamoeba fragilis*: an emerging role in intestinal disease. CMAJ 175(5):468–469

Lanuza MD, Carbajal JA, Villar J, Mir A, Borras R (1999) Soluble-protein and antigenic heterogeneity in axenic *Blastocystis hominis* isolates: pathogenic implications. Parasitol Res 85(2):93–97

Leelayoova S, Siripattanapipong S, Thathaisong U, Naaglor T, Taamasri P, Piyaraj P, Mungthin M (2008) Drinking water: a possible source of *Blastocystis* spp. subtype 1 infection in schoolchildren of a rural community in central Thailand. Am J Trop Med Hyg 79(3):401–406

Leelayoova S, Taamasri P, Rangsin R, Naaglor T, Thathaisong U, Mungthin M (2002) In-vitro cultivation: a sensitive method for detecting *Blastocystis hominis*. Ann Trop Med Parasitol 96 (8):803–807

Li LH, Zhang XP, Lv S, Zhang L, Yoshikawa H, Wu Z, Steinmann P, Utzinger J, Tong XM, Chen SH, Zhou XN (2007a) Cross-sectional surveys and subtype classification of human *Blastocystis* isolates from four epidemiological settings in China. Parasitol Res 102(1):83–90

Li LH, Zhou XN, Du ZW, Wang XZ, Wang LB, Jiang JY, Yoshikawa H, Steinmann P, Utzinger J, Wu Z, Chen JX, Chen SH, Zhang L (2007b) Molecular epidemiology of human *Blastocystis* in a village in Yunnan province, China. Parasitol Int 56(4):281–286

Mahmoud MS, Saleh WA (2003) Secretory and humoral antibody responses to *Blastocystis hominis* in symptomatic and asymptomatic human infections. J Egypt Soc Parasitol 33 (1):13–30

Markell EK, Havens RF, Kuritsubo RA (1983) Intestinal parasitic infections in homosexual men at a San Francisco health fair. West J Med 139(2):177–178

Markell EK, Havens RF, Kuritsubo RA, Wingerd J (1984) Intestinal protozoa in homosexual men of the San Francisco Bay area: prevalence and correlates of infection. Am J Trop Med Hyg 33 (2):239–245

Markell EK, Udkow MP (1986) *Blastocystis hominis*: pathogen or fellow traveler? Am J Trop Med Hyg 35(5):1023–1026

Markell EK, Voge M (1976) Medical parasitology. WB Saunders Company, Philadelphia, PA

Matukaitis JM (1997) The emerging recognition of *Cryptosporidium* as a health hazard. J Community Health Nurs 14(3):135–140

Mirza H, Teo JD, Upcroft J, Tan KS (2010) A rapid, high-throughput viability assay for *Blastocystis* spp. reveals metronidazole resistance and extensive subtype-dependent variations in drug susceptibilities. Antimicrob Agents Chemother 55(2):637–648

Mirza H, Wu Z, Kidwai F, Tan KS (2011) A metronidazole-resistant isolate of *Blastocystis* spp. is susceptible to nitric oxide and downregulates intestinal epithelial inducible nitric oxide synthase by a novel parasite survival mechanism. Infect Immun 79(12):5019–5026

Moe KT, Singh M, Howe J, Ho LC, Tan SW, Chen XQ, Ng GC, Yap EH (1997) Experimental *Blastocystis hominis* infection in laboratory mice. Parasitol Res 83(4):319–325

Morrison HG, McArthur AG, Gillin FD, Aley SB, Adam RD, Olsen GJ, Best AA, Cande WZ, Chen F, Cipriano MJ, Davids BJ, Dawson SC, Elmendorf HG, Hehl AB, Holder ME, Huse SM, Kim UU, Lasek-Nesselquist E, Manning G, Nigam A, Nixon JE, Palm D, Passamaneck NE, Prabhu A, Reich CI, Reiner DS, Samuelson J, Svard SG, Sogin ML (2007) Genomic minimalism in the early diverging intestinal parasite *Giardia lamblia*. Science 317(5846):1921–1926

Nigro L, Larocca L, Massarelli L, Patamia I, Minniti S, Palermo F, Cacopardo B (2003) A placebo-controlled treatment trial of *Blastocystis hominis* infection with metronidazole. J Travel Med 10(2):128–130

Nimri LF (1993) Evidence of an epidemic of *Blastocystis hominis* infections in preschool children in northern Jordan. J Clin Microbiol 31(10):2706–2708

Noel C, Dufernez F, Gerbod D, Edgcomb VP, Delgado-Viscogliosi P, Ho LC, Singh M, Wintjens R, Sogin ML, Capron M, Pierce R, Zenner L, Viscogliosi E (2005) Molecular phylogenies of *Blastocystis* isolates from different hosts: implications for genetic diversity, identification of species, and zoonosis. J Clin Microbiol 43(1):348–355

O'Donoghue PJ (1985) *Cryptosporidium* infections in man, animals, birds and fish. Aust Vet J 62 (8):253–258

Olivo-Diaz A, Romero-Valdovinos M, Gudino-Ramirez A, Reyes-Gordillo J, Jimenez-Gonzalez DE, Ramirez-Miranda ME, Martinez-Flores WA, Martinez-Hernandez F, Flisser A, Maravilla P (2012) Findings related to IL-8 and IL-10 gene polymorphisms in a Mexican patient population with irritable bowel syndrome infected with *Blastocystis*. Parasitol Res 111 (1):487–491

Parkar U, Traub RJ, Vitali S, Elliot A, Levecke B, Robertson I, Geurden T, Steele J, Drake B, Thompson RC (2010) Molecular characterization of *Blastocystis* isolates from zoo animals and their animal-keepers. Vet Parasitol 169(1–2):8–17
Pickering LK, Woodward WE, DuPont HL, Sullivan P (1984) Occurrence of *Giardia lamblia* in children in day care centers. J Pediatr 104(4):522–526
Rosenblatt JE (1990) *Blastocystis hominis*. J Clin Microbiol 28(10):2379–2380
Santin M, Gomez-Munoz MT, Solano-Aguilar G, Fayer R (2011) Development of a new PCR protocol to detect and subtype *Blastocystis* spp. from humans and animals. Parasitol Res 109 (1):205–212
Silard R, Burghelea B (1985) Ultrastructural aspects of *Blastocystis hominis* strain resistant to antiprotozoal drugs. Arch Roum Pathol Exp Microbiol 44(1):73–85
Smith JW, Wolfe MS (1980) Giardiasis. Annu Rev Med 31:373–383
Souppart L, Moussa H, Cian A, Sanciu G, Poirier P, El Alaoui H, Delbac F, Boorom K, Delhaes L, Dei-Cas E, Viscogliosi E (2010) Subtype analysis of *Blastocystis* isolates from symptomatic patients in Egypt. Parasitol Res 106(2):505–511
Stark D, Barratt J, Roberts T, Marriott D, Harkness J, Ellis J (2012) A review of the clinical presentation of dientamoebiasis. Am J Trop Med Hyg 82(4):614–619
Stark D, Hal van S, Marriott D, Ellis J, Harkness J (2007) Irritable bowel syndrome: a review on the role of intestinal protozoa and the importance of their detection and diagnosis. Int J Parasitol 37(1):11–20
Stensvold CR, Arendrup MC, Jespersgaard C, Molbak K, Nielsen HV (2007a) Detecting *Blastocystis* using parasitologic and DNA-based methods: a comparative study. Diagn Microbiol Infect Dis 59(3):303–307
Stensvold CR, Christiansen DB, Olsen KE, Nielsen HV (2011) *Blastocystis* sp. subtype 4 is common in Danish *Blastocystis*-positive patients presenting with acute diarrhea. Am J Trop Med Hyg 84(6):883–885
Stensvold CR, Smith HV, Nagel R, Olsen KE, Traub RJ (2010) Eradication of *Blastocystis* carriage with antimicrobials: reality or delusion? J Clin Gastroenterol 44(2):85–90
Stensvold CR, Suresh GK, Tan KS, Thompson RC, Traub RJ, Viscogliosi E, Yoshikawa H, Clark CG (2007b) Terminology for *Blastocystis* subtypes–a consensus. Trends Parasitol 23(3):93–96
Stenzel DJ, Boreham PF (1996) *Blastocystis hominis* revisited. Clin Microbiol Rev 9(4):563–584
Swan R (1983) Faith healing, Christian Science, and the medical care of children. N Engl J Med 309(26):1639–1641
Tan KS, Mirza H, Teo JD, Wu B, Macary PA (2011) Current views on the clinical relevance of *Blastocystis* spp. Curr Infect Dis Rep 12(1):28–35
Tan TC, Ong SC, Suresh KG (2009) Genetic variability of *Blastocystis* sp. isolates obtained from cancer and HIV/AIDS patients. Parasitol Res 105(5):1283–1286
Udkow MP, Markell EK (1993) *Blastocystis hominis*: prevalence in asymptomatic versus symptomatic hosts. J Infect Dis 168(1):242–244
US_EPA (1979) Proceedings of the EPA Symposium on giardiasis. http://nepis.epa.gov/Exe/ZyPURL.cgi?Dockey=2000UCBL.txt.179
Walters LR (2004) Human embryonic stem cell research: an intercultural perspective. Kennedy Inst Ethics J 14(1):3–38
Windsor JJ (2007) *B. hominis* and *D. fragilis*: Neglected human protozoa. The Biomedical Scientist: 51(7):524–527
Windsor JJ, Macfarlane L (2005) Irritable bowel syndrome: the need to exclude *Dientamoeba fragilis*. Am J Trop Med Hyg 72(5):501, author reply 501-2
Xu P, Widmer G, Wang Y, Ozaki LS, Alves JM, Serrano MG, Puiu D, Manque P, Akiyoshi D, Mackey AJ, Pearson WR, Dear PH, Bankier AT, Peterson DL, Abrahamsen MS, Kapur V, Tzipori S, Buck GA (2004) The genome of *Cryptosporidium hominis*. Nature 431 (7012):1107–1112

Yamamoto-Furusho JK, Torijano-Carrera E (2010) Intestinal protozoa infections among patients with ulcerative colitis: prevalence and impact on clinical disease course. Digestion 82 (1):18–23

Yao FR, Qiao JY, Zhao Y, Zhang X, Yang JH, Li XQ (2005) Experimental infection of mice with *Blastocystis hominis*. Zhongguo Ji Sheng Chong Xue Yu Ji Sheng Chong Bing Za Zhi 23 (6):444–448

Yoshikawa H, Dogruman-Al F, Turk S, Kustimur S, Balaban N, Sultan N (2011) Evaluation of DNA extraction kits for molecular diagnosis of human *Blastocystis* subtypes from fecal samples. Parasitol Res 109(4):1045–1050

Zaman V, Zaki M (1996) Resistance of *Blastocystis hominis* cysts to metronidazole. Trop Med Int Health 1(5):677–678

Zheng H, Jia H, Shankar A, Heneine W, Switzer WM (2011) Detection of murine leukemia virus or mouse DNA in commercial RT-PCR reagents and human DNAs. PLoS One 6(12):e29050

Zierdt CH (1991a) *Blastocystis hominis*–past and future. Clin Microbiol Rev 4(1):61–79

Zierdt CH (1991b) Pathogenicity of *Blastocystis hominis*. J Clin Microbiol 29(3):662–663

Zierdt CH, Zierdt WS, Nagy B (1995) Enzyme-linked immunosorbent assay for detection of serum antibody to *Blastocystis hominis* in symptomatic infections. J Parasitol 81(1):127–129

Chapter 9
Diarrheas in Humans Due to Agents of Disease

Heinz Mehlhorn

Abstract *Blastocystis* species and related species—the topics of this book—introduce intestinal symptoms of diseases that are described in literature as diarrheas of dysenterias. Thus this short article intends to present a condensed survey on other agents of diseases leading to similar symptoms as infections with *Blastocystis* species. In general these symptoms are not very specific, so that intense diagnosis must be done in order to find out the responsible agents of disease and thus to have the chance to apply the most helpful measurements of therapy and prophylaxis.

9.1 Introduction

The target organisms of this volume of the book series Parasitology Research Monographs are human or animal *Blastocystis* species, which may lead under certain, but not completely understood conditions to severe diarrheas (Darai et al. 2009; Löscher and Burchard 2010; Mehlhorn 2008, 2012a, b; Neumeister et al. 2009; Mertens et al. 2004).

This uptake of agents of disease may occur on the oral pathway after direct contact with feces or drinking of contaminated water respectively after eating contaminated food. However, such agents of disease may also become transmitted indirectly by flies as was recently proven by investigating the bodies, feces, feet, and mouthparts of flies caught in the nature close to feces of dogs, cattle, rabbits, chicken, pigs, and horses (Gestmann et al. 2012; Förster et al. 2012). These findings in nature were confirmed by the same group of researchers when exposing flies to defined amounts of parasites and bacteria and controlling the transmission potential. Among the detected pathogens a broad spectrum of bacteria and parasites was

H. Mehlhorn (✉)
Institute for Parasitology, Heinrich Heine University, Universitätsstr. 1
D-40225 Düsseldorf, Germany
e-mail: mehlhorn@uni-duesseldorf.de

H. Mehlhorn et al. (eds.), *Blastocystis: Pathogen or Passenger?*, 165
Parasitology Research Monographs 4,
DOI 10.1007/978-3-642-32738-4_9, © Springer-Verlag Berlin Heidelberg 2012

isolated, which are known to induce clinical symptoms of disease that in common are described as diarrhea respectively dysenteria. However, all symptoms observed are in general rather similar so that it is needed to start microscopical and/or molecular biological test series in order to diagnose the exact agent of disease or to find explanations pointing to an organic disease of the intestinal tract. Otherwise chemotherapy—wherever possible—could not be accomplished or dietic measurements would be the wrong decision in severe cases of infections with bacteria. Only the correct diagnosis will finally allow to introduce additional sanitary measurements which could reduce the risk of a mass infection in a group of humans living close together.

Thus it is worthwhile—especially under the aspects of a necessary differential diagnosis—to throw a glimpse onto other factors respectively other agents (viruses, bacteria, fungi, parasites) that may introduce diarrheal stools in humans leading in many cases to life threatening diseases due to an often extreme loss of water (in some cases being accompanied by blood loss). Such diseases and their primarily gastro-intestinally based symptoms may have their origin in inappropriate food or in physiological defects of the intestine but also may occur as a consequence of the uptake of agents of diseases.

Especially in malnourished persons living closely together in camps with bad hygienic conditions local outbreaks may lead to intense rushes of transmissions of such agents, which may belong either to the groups of viruses, fungi, bacteria, or parasites. Since in many cases chemotherapy is not possible at all or not available at the local sites of outbreaks, a quick stop of such epidemics is mostly not possible and thus leads to a spreading of the agents of diseases by travelers, goods, and/or insect vectors over long distances even into countries with high standards of hygiene.

9.2 Definitions and Pathogens

Diarrhea (*Greek*: dia = through; rhein = flowing) and dysenteria (*Greek*: dys = bad function; enteron = intestine) describe diseases which are characterized respectively defined by the fact that concerned persons excrete liquid feces at least more often than three times per day and thus are endangered by a considerable loss of water and electrolytic compounds, which then may introduce a potential collapse of the body especially in severe (=acute) cases, where large amounts of mucus, pus, and/or blood are additionally included therein.

While the terms diarrhea or dysenteria cover in a more general sense, practically all malfunctions of the intestinal tractus lead to increased liquid defecations. The terms gastroenteritis (*Greek*: gaster = stomach; enteron = intestine; -itis = malfunction), cholangitis (*Greek*: chole = bile, gall), and colitis (*Greek*: kolon = terminal portion of the intestine prior to rectum and anus) are used to describe regional intestinal diseases of the stomach, bile producing system respectively of the mid or terminal portions of the intestinal tract. With respect to the infectious intestinal diseases there are many agents of diseases that are able to introduce rather similar appearing symptoms of disease (Tables 9.1, 9.2, 9.3) (Darai et al. 2009; Kist 2009; Reuter 2004; Streckert 2009a, b, c; Tschäpe et al. 2009).

The various clinical symptoms of diarrheas and dysenterias are the basis of the description of different subgroups of these diseases which often show overlapping symptoms.

9.2.1 Diarrheas

1. *Acute or catarrhalic diarrhea*: with numerous, liquid defecations per day.
2. *Chronic diarrhea*: with long lasting symptoms.
3. *Lientric diarrhea*: with indigested food particles in stool.
4. *Profuse diarrhea*: strong, watery stools.
5. *Malassimilatory diarrhea*: food is resorbed.
6. *Mucus diarrhea*: stool contains large amounts of slimy elements.
7. *Diarrhea paradoxa*: as follow up of an obstipation.
8. *Tropical diarrhea*: especially disease of travelers due to different agents of disease.
9. *Inflammatory diarrhea* (Table 9.4): is based on invasion and destruction of the intestinal wall by the pathogen and/or excreted toxins.
10. Noninflammatory diarrhea (Table 9.5): numerous watery defecations, often with convulsions.

9.2.2 Dysenterias

Many authors claim for the symptoms of this group, that bloody components are essential diagnostic tools.

1. *Dysenteria endemica*: due to *Entamoeba histolytica* (intestinal phase).
2. *Dysenteria hepatica*: due to *Entamoeba histolytica* (liver abscess).
3. *Dysenteria fulminans*: stormy with heavy water loss.
4. *Dysenteria maligna*: life threatening, often deadly form mostly due to bacterial invasion in tissues.

9.3 Importance and Pathways of Infections of Diarrheas

Diarrheas respectively dysenterias are very important human infectious diseases which reached in times of climatic changes and intense globalization worldwide the status of *emerging diseases* (Mehlhorn 2008; Mehlhorn 2012a, b, c). The transmission of the pathogens occurs exclusively orally—fecally by oral

Table 9.1 Alphabetical list of pathogens that may induce diarrheas (selection)

Agent of disease (genus/species)	Characteristics and systematical position of the pathogen
Arcobacter butzleri, etc.	Gram-negative, spirallic bacteria; Campylobacteriaceae
Balantidium coli	Parasite: Protozoa, Ciliate
Blastocystis hominis	Parasite: Globular unicellular organism with varying size of 5–200 μm in diameter
Caliciviren	See Norwalk-Virus: small RNA-viruses of 26–35 nm in size without capsules, icosaedric, cup-like shape
Campylobacter jejuni and other	Gram-negative spirallic bacteria; Campylobacteriaceae, Protobacteria
Citrobacter freundii	Gram-negative, rod-like bacteria with endotoxins; Enterobacteriaceae
Clostridium perfringens, etc.	Gram-positive, anaerobic, short rod-like bacteria (with subterminal spore); Bacillaceae
Coxsacki virus	Human enterovirus; Picornaviridae; spherical, not covered RNA-virus of 30 nm in diameter
Cryptosporidium species	Parasites: Protozoa; spherical oocysts with 5 μm in diameter
Cyclospora cayetanensis	Parasite: Protozoa; Apicomplexa; oocysts with two sporocysts each with two sporozoites
Cytomegalovirus (HCMV)	Human herpes virus linear, double strand DNA-virus with capsid; Herpesviridae
Dicrocoelium dendriticum	Parasite: trematode lives in bile ducts, up to 15 mm in length
Dientamoeba fragilis	Parasite: amoeba-like protozoan of up to 12 μm, may form cysts
Echovirus, *Parechovirus*	Human enterovirus, spherical RNA virus of 30 nm in diameter, without cover
Entamoeba histolytica	Parasite: Amoeba: Minuta- (10–20 μm) and Magna-form (40 μm). The latter may enter intestinal wall and liver forming abscesses
Enterobacter sakazakii, *E. aerogenes*, etc.	Gram-negative rod-like bacteria of up to 3 μm in length, formation of capsules and endotoxin; Enterobacteriaceae
Escherichia coli and its variations (EHEC, ETEC, EIEC, EPEC, AEEC)	Gram-negative, rod-like bacteria of 1.5 × 2–6 μm in size, with or without capsules. EPEC, etc. are obligate pathogens; Enterobacteriaceae
Giardia lamblia (syn. *duodenalis*) and other assemblages (A–G)	Parasites: flagellated protozoan (with 8 flagella as trophozoite; cyst with 4 nuclei); size 10–20 μm
Hantavirus (*Hantaan*-virus, etc.)	Virus with a minus single strand RNA genome with a diameter of 100 nm containing three separate nucleocapsids
Hepatitis-A-virus (HAV)	Picornaviridae-virus of 27 nm in diameter without cover containing a single strand linear+, RNA genome of about 7.5 kb in length
Hookworms (genera *Ancylostoma*, *Necator*)	Parasites: nematode worms of about 1–1.5 cm in length, suck blood in both sexes at the intestinal wall

(continued)

Table 9.1 (continued)

Agent of disease (genus/species)	Characteristics and systematical position of the pathogen
Hymenolepis nana, etc.	Parasite: Tapeworms of 5–6 cm in length, self-infections possible = mass infection.
Isospora belli (syn. *hominis*)	Parasite: Protozoan: Apicomplexa; forms oocysts of 20–30 μm × 10–20 μm containing two sporocysts each with 4 sporozoites
Liver flukes (*Opisthorchis*, *Clonorchis*)	Parasites: Adult trematode worms of about 1.5 cm in length live in the bile ducts; infection by oral uptake of infected fis.
Legionella species	Gamma protobacteria; gram-negative, facultatively intracellular, not sporeforming, short coccoid rod-like bacteria of 0.3–0.9 μm × 1.5–2 μm in size
Leishmania donovani, etc.	Parasite: singly flagellated (short, hidden flagellum) protozoan that reproduces in humans intracellularly; transmitted by bites of sandflies
Listeria monocytogenes, *L. ivanovii*	Gram-positive, mobile bacteria, not spore forming rod-like short bacteria with facultative intracellular reproduction; Lactobacillidae
Measles-virus, *Morbilli*-virus	Virion has a size of 110–250 nm and has a cover surrounding the helical nucleocapsid with a RNA-genome; Paramyxoviridae
Microsporidia (e.g., *Encephalitozoon cuniculi*, *Enterocytozoon bieneusi*)	Parasites small-sized, intracellular protozoans (2.5 μm), form cysts in immune-suppressed persons; generalization as opportunistic agents
Neisseria meningitides	Gram-negative, diplococcoid, non-motile, non-spore forming, aerobic β-proterobacteria of 1 μm in diameter
Norwalk-(Noro)virus (NV)	Non-covered, 27 nm sized, icosaedric virus of 27 nm in diameter containing a RNA-genome, Caliciviridae
Plesiomonas sp. (syn. *P. shigelloides*, *Aeromonas shigelloides*)	Motile gram-negative, rod-like bacteria with flagella at one of the two cell poles, Enterobacteriaceae
Rotavirus (6 human serotypes; 2 groups)	Virus of the family Reoviridae with a diameter of about 70 nm containing 11 segments of double-strand RNA
Salmonella enterica, etc.	Bacteria of the family Enterobacteriaceae; gram-negative, capsule-less, non-spore forming, rod-like bacteria with a size of 1.3–4 × 2–6 μm
Sarcocystis species (*S. suihominis*, *S. bovihominis*)	Parasites: Protozoa, Apicomplexa. Humans excrete oocysts with two sporocysts and each with four sporozoites or single sporocysts (after disruption of oocyst wall)
Shigella dysenteriae	Bacteria of the family Enterobacteriaceae; gram-negative, aerobic, rod-like bacteria without spore formation with a size of 2–6 μm × 1.3 μm; they invade intestinal cells; *E. dysenteriae* produces Shigatoxin; some authors consider *Shigella* as *E. coli*-pathovars

(continued)

Table 9.1 (continued)

Agent of disease (genus/species)	Characteristics and systematical position of the pathogen
Staphylococcus aureus	Bacteria of the new classified family Micrococcaceae; stages are gram-positive coccoid bacteria, which often form clusters during growth
Strongyloides stercoralis	Parasite: nematode worm, which lives as parthenogenic, 2 mm long females in human intestine and as bisexual generation outside of the body (reaching lengths of 0.8–1 mm)
Taenia species (e.g., *T. solium, T. saginata, T. asiatica*)	Parasites: tapeworms, reaching lengths of up to 8 m in human intestine; mostly only one adult worm per person
Trichinella spiralis and related species	Parasites: nematode worms: ♀, ♂ of ~1–3 mm live in human intestine; (100 μm long larvae live intracellularly in human muscle cells)
Trichuris trichiura	Parasite: nematode worms of 5–6 cm in length, live in the colon of humans; their thin anterior end penetrates into the interior of intestinal cells
Vibrio cholerae, etc.	Bacteria belonging to the family Vibrionaceae; stages are 0.5–0.8 μm × 1.4–2.6 μm sized vibrions with comma-appearance and gram-negative reaction, polar flagella
Yersinia enterocolitica, etc.	Proterobacteria (Fam. Enterobacteriaceae), short rod-like, gram-negative bacteria

uptake of persistent stages (cysts, spores, etc.) of the various pathogens (Tables 9.1, 9.2, 9.3, 9.4, 9.5) from human or animal feces by direct contact or by uptake of contaminated food and/or drinking water. The food or human mouthparts may also become contaminated indirectly by help of licking flies. Many studies (literature, see Förster et al. 2012; Gestmann et al. 2012) showed that the body, the mouthparts, and/or feet of numerous fly species may carry large amounts of pathogens (Figs. 9.1, 9.2, 9.3).

Furthermore the feces of blood-sucking insects may contain agents of diseases, which then might become inhaled or ingested by humans. The experiments of Mencke et al. (2009) and Vobis et al. (2003, 2005) clearly showed that feces of cat fleas, which had taken up blood from animals being infected with *Fe*line *L*ymphoma *V*irus (FeLV) or *Fe*line *C*alici-*V*irus (FeCV) remain infectious for weeks. Thus virus-infected flea feces may endanger for weeks in humans who get into contact with pathogens containing feces of fleas.

Since in most cases only a few specimens of pathogens are needed to establish a progressive infection, the transmission pathway by flies and fleas is probably worldwide still underestimated.

According to reasonable calculations of WHO (2004) and other groups Modis (2012) ranges human diarrheal infections on place two after parasitic diseases asking 72.8 million DALYs (=*D*iasability *A*djusted *L*ife *Y*ears; Murray 1994; Murray and Lopez 1996; Mathers 2008) and thus range significantly before AIDS

Table 9.2 Alphabetical list of pathogens that may induce gastroenteritis

Agent of disease (genus/species)	Characteristics and systematical position of the agent of disease
Adenovirus	Family Adenoviridae: virus (without cover), measures 70–90 nm size and contains a linear, double-strand DNA-genome
Astrovirus	Family Astroviridae (genus Mamastrovirus); the 27–30 nm sized, icosaedric capsid contains a single strand RNA-genome
Blastocystis hominis	See Table 9.1
Campylobacter species	See Table 9.1
Clostridium species	See Table 9.1
Coxsackivirus	See Table 9.1
Echovirus, Paraechovirus	See Table 9.1
Edwardsiella sp.	Gram-negative, rod-like, motile bacteria with a size of 1 × 2–3 μm
Enterobacter sp.	See Table 9.1
Escherichia coli	See Table 9.1
Listeria monocytogenes	See Table 9.1
Norwalk-(Noro-)*virus*	See Table 9.1
Reovirus (*Orthovirus*)	See Table 9.1
Rotavirus	See Table 9.1
Salmonella enterica, etc.	Bacteria of the family Enterobacteriaceae; gram-negative, capsule-less, non-spore forming, rod-like bacteria with a size of 1.3–4 × 2–6 μm
Shigella sp.	See Table 9.1
Trichinella species	See Table 9.1
Vibrio species	See Table 9.1
Yersinia enterocolitica, etc.	Proterobacteria (Fam. Enterobacteriaceae), short rod-like, gram-negative bacteria

Table 9.3 Alphabetical list of pathogens that may induce colitis

Agent of disease (genus/species)	Characteristics and systematical position of the agent of disease
Campylobacter jejuni	See Table 9.1
Clostridium species	See Table 9.1; *C. difficile* after use of antibiotica; endotoxins attack intestinal walls and induce necrosis
Cytomegalovirus	Family Herpesviridae; virus possesses icosaedric capsid, tegument and outer membrane. Genome: double strand DNA
Entamoeba histolytica	See Table 9.1
Escherichia coli	See Table 9.1

(58.5 DALYs), tuberculosis (34.2 DALYs), or malaria (34 DALYs). In total it is estimated that about 20 % of the diarrheal diseases are induced by transmitted parasites and the rest by various viral and bacterial pathogens.

However, the sources of outbreaks of diarrheas respectively dysenterias remain often hidden due to the lack of appropriate methods for diagnosis (Streckert 2009a, b, c). Thus documentating reviews on outbreaks of diarrheal diseases show often the pattern that was described by Barnes et al. (1998) in Australia for 4,637 hospitalized children. This group documented that in 43.3 % of the cases, the agents

Table 9.4 Pathogens inducing an inflammatory diarrhea (selection)

Species	Incubation period	Persistence of symptoms	Fever > 38.5 °C
Campylobacter species	6–48 h	~7 d	+
Entamoeba histolytica	2 d–3 months	Diarrhea: 6–14 d Abscess: months	+ (rare)
Enteric *Salmonella* species	6–48 h	1–10 w	+
Enterohaemorrhagic *Escherichia coli* (EHEC, etc.)	1–8 d	3–7 d	+ (rare)
Typhic *Salmonella* species	1–3 w	~4 w	+
Vibrio parahaemolyticus	5 h–1 d	~7–10 d	+
Yersina species	10–48 h	1–3 w	+

>: higher, more than; +: yes; ~: about; d: days; h: hours; m: months; w = weeks

Table 9.5 Pathogens inducing a non-inflammatory diarrhea (selection)

Species	Incubation period	Persistence of symptoms	Fever > 38.5 °C
Blastocystis hominis	2–3 d–3 w	~5 w	+ (rare)
Clostridium perfringens	14–24 h	~24 h	+ (rare)
Cryptosporidium species	1–2 d	3–12 d	–
Cyclospora cayetanensis	2–7 d	2–12 w	–
Enterotoxic *Escherichia coli* (ETEC)	16–72 h	~10 d	+ (rare)
Giardia lamblia	~1–2 w	~months	–
Isospora belli	2–13 d	2–4 w–1 year	–
Norwalk-Virus	1–2 d	1–2 d	+ (rare)
Rotavirus	1–3 d	6–8 d	+
Sarcocystis suihominis	4–6 h	~24–48 h	–
Vibrio cholerae	1–3 d	6–9 d	–

+: present; –: not present; ~: about; d: days; h: hours; w: weeks

Fig. 9.1 Macrophoto of a *Musca* fly in its typical position of licking at a surface

of the gastroenteritis was not found, while 39.6 % were diagnosed to occur as consequence of a rotavirus infection, 6 % respectively 1.9 % as follow ups of infections with specimens of the adenovirus—or astrovirus group. Only 5.8 % were due infections with bacteria of the *Salmonella* group respectively 3.4 % due to species of the *Campylobacter* group.

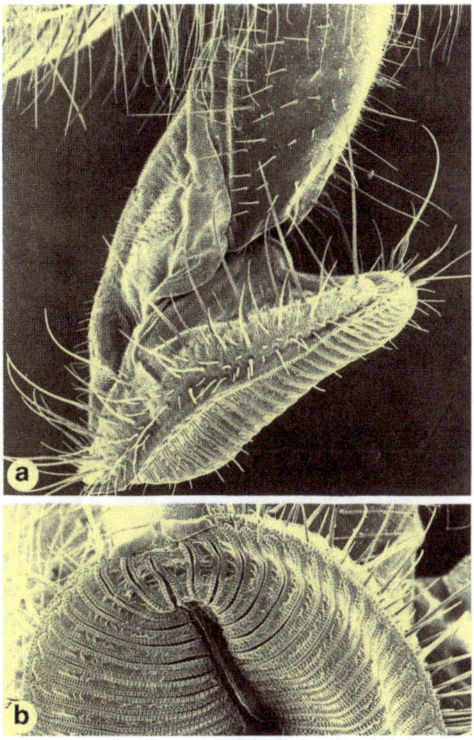

Fig. 9.2 Scanning electron micrographs of the mouthparts of a *Musca* fly showing the enlarged tips of the labellae, which are provided with numerous grooves that are filled with gluing saliva. (**a**) Lateral view, (**b**) ventral surface of the licking labellae

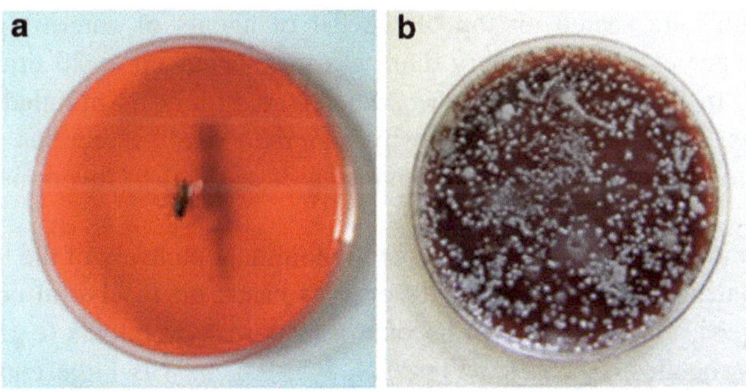

Fig. 9.3 Petri dish with bacteria growing medium when just exposed to a *Musca* fly (**a**) and after 2 days with fully grown bacterial colonies (**b**)

With respect to the death toll, rotaviral infections are forerunners killing up to 650,000 persons per year (especially children of 6 months until 2 years in age). 90 % of these deaths occur in Asia and Africa (Glass et al. 2005). Also the species of the adenovirus group lead to disseminated diseases with a lethality of 20–80 %,

Fig. 9.4 Macrophoto of an American cockroach (*Periplaneta americana*) It came in the 17th century with sugar canes from USA to Europe

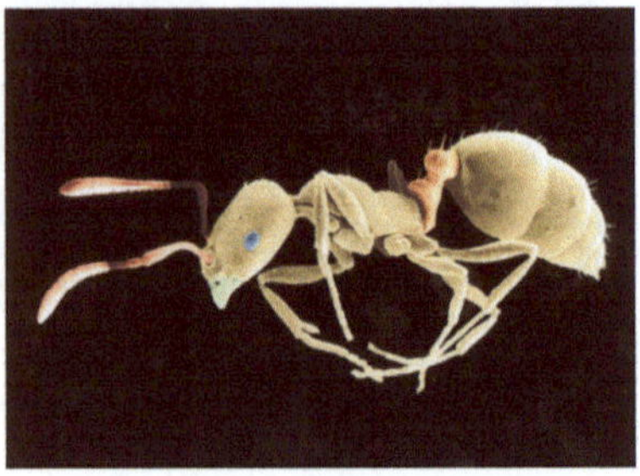

Fig. 9.5 Scanning electron micrograph of a so-called pharao ant (*Monomorium pharaonis*), which like to feed on human protein food and on wounds

while the obligatorily notifiable disease due to *Salmonella* leads only in 5 % of the cases to enteritic symptoms (Tschäpe et al. 2009). In the case of infections with *Campylobacter* species (*C. jejuni*; *C. coli*) it turned out that their number increases worldwide and are seated on top of the list of agents of enteritis prior to the *Salmonella* species. Acute, watery diarrheas may occur up to 20 times per 24 h (Kist 2009). In 30 % of the *Campylobacter* stools blood is mingled among the watery feces. This makes it clear why *Campylobacter* infections have to be announced in many countries as notifiable diseases to the Governmental Health Authorities.

Although it is absolutely clear that the predominant pathway of the transmission of agents diarrheas respectively dysenterias is based on fecal–oral contacts with such pathogens, their potential to become transported by insects (e.g., ants, flies, cockroaches, beetles, fleas, etc., Figs. 9.4, 9.5, 9.6, 9.7) is large enough that in communities of closely "packed" people, severe and wide epidemics may become initiated—especially in those cases where the sanitary conditions are poor (Figs. 9.8 and 9.9).

All literature on diarrheal diseases clearly points out, that in all cases the severeness of the diseases depends largely on the fitness of the immune system of the persons that become infected. Kids below 2 years and immuno-compromised

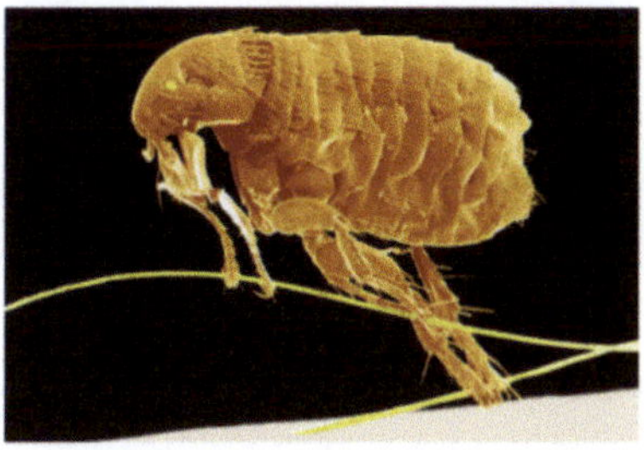

Fig. 9.6 Scanning electron micrograph of an adult cat flea (*Ctenocephalides felis*)

Fig. 9.7 Macrophoto of a larva of the flour beetle (*Tenebrio molitor*), which is often found on human food

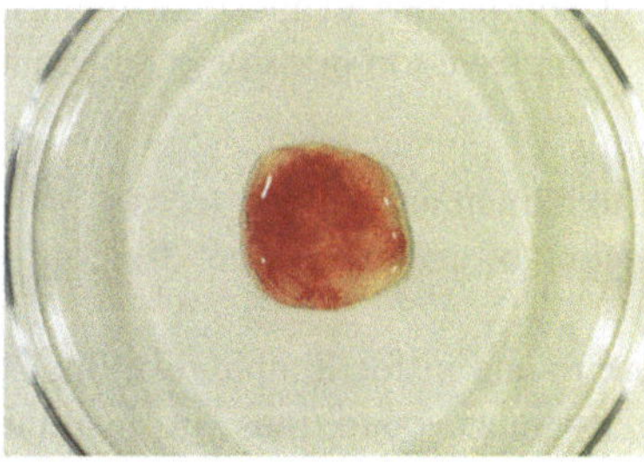

Fig. 9.8 Fluid feces with bloody contents after an infection with the amoeba *Entamoeba histolytica*

persons are therefore most highly endangered. In these groups the severeness of the clinical symptoms reaches the highest values and even considerable rates of lethality (Neumeister et al. 2009; Löscher and Burchard 2011; Darai et al. 2009; Mehlhorn 2012a, b, c).

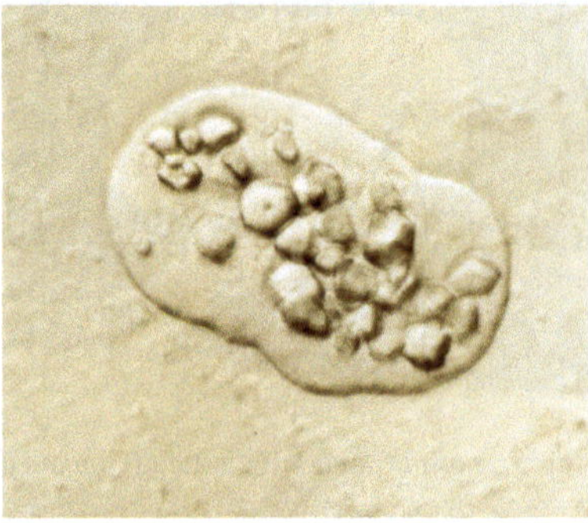

Fig. 9.9 Light micrograph of a so-called magna-stage of *Entamoeba histolytica* that is able to penetrate into the intestinal wall of humans (it contains already red blood cells)

9.4 Conclusions

Diarrheas and dysenterias are worldwide very wide spread and may lead to severe life-threatening diseases. The situation becomes worse when considering that in many types of diseases the means and methods of diagnosis are in general not existing, poor, or not available in many countries. Furthermore there are practically no remedies to cure viral diseases completely. Furthermore many bacterial pathogens have developed considerable resistances against several groups of antibiotics. Thus the recent situation is worldwide in a critical phase and not only in countries with low hygienic standards.

Therefore as important approaches in the fight against diarrheal diseases the following measurements must have top priority in all countries wherever on earth:

- Control of fecal deposals
- Avoidance of contact to human and animal feces
- Production and provision of pathogen-free drinking water for humans and animals
- Control of insects close to and inside human dwellings and animal stables
- Constant control of hygienic conditions in institutions, food production, restaurants, shops, etc.
- Establishment of an alarm system to recognize epidemics as early as possible
- Repeated, reasonable information of the public on the pathways of transmission of pathogens and available methods of prophylaxis

References

Barnes GL, Uren E, Stevens KB et al (1998) Etiology of acute gastroenteritis in hospitalized children in Melbourne, Australia. J Clin Microbiol 36:133–138

Darai G, Handermann M, Sonntag HG, Tidona CA, Zöller L (2009) Lexikon der Infektionskrankheiten des Menschen, 3rd edn. Springer, Heidelberg

Förster M, Gestmann F, Mehlhorn H, Sievert K, Messler S, Neuhausen N, Petersdorf S, Pfeffer K (2012) In: Mehlhorn H (ed) Flies as vectors of parasites potentially inducing severe diseases, vol 3, Parasitology research monographs. Springer, Heidelberg, New York

Gestmann F, Förster M, Mehlhorn H, Messler S, Neuhausen N, Pfeffer K (2012) Flies as vectors of parasites potentially inducing severe diseases in humans and animals. In: Mehlhorn H (ed) Arthropod as vectors of emerging diseases, vol 3, Parasitology research monographs. Springer, Heidelberg

Glass RI, Breese JS, Turcios R et al (2005) Rotavirus vaccines: targeting the developing world. J Infect Dis 192:160–166

Kist M (2009) *Campylobacter* and *Arcobacter*. In: Neumeister B et al (eds) Microbiological diagnostic, 2nd edn. Thieme, Stuttgart

Löscher T, Burchard RD (2010) Reisemedizin, 2nd edn. Thieme, Stuttgart

Mathers G (2008) The global burden of disease: 2004 update. WHO, Geneva

Mehlhorn H (ed) (2008) Encyclopedia of parasitology, 3rd edn. Springer, New York, Heidelberg

Mehlhorn H (2012a) The parasites of animals, 7th edn. Springer, Spektrum Akad. Verlag, Heidelberg

Mehlhorn H (2012b) The parasites of man and the diseases, 7th edn. Springer, Spektrum Akad. Verlag, Heidelberg

Mehlhorn H (ed) (2012c) Arthropods as vectors of emerging diseases. Parasitology research monographs, vol 3, Springer, Heidelberg

Mencke N, Vobis M, Mehlhorn H, D'Haese J, Rehagen M, Mangold-Gehring S, Truyen U (2009) Transmission of the feline calicivirus via the cat flea. Parasitol Res 105:185–189

Mertens T, Haller O, Klenk HD (2004) Diagnostik und Therapie von Viruskrankheiten, 2nd edn. Elsevier, München

Modis Y (2012) Exploiting structural biology in the fight against parasitic diseases. Trends Parasitol 28:124–130

Murray CJ (1994) Quantifying the burden of disease: the technical basis for disability-adjusted-life years. Bull World Health Org 72:429–449

Murray CJ, Lopez AD (1996) Evidence-based health policy. Science 274:740–743

Neumeister B, Geiss KH, Braun RW, Kimmig P (eds) (2009) Microbiological diagnostic, 2nd edn. Thieme, Stuttgart

Reuter P (ed) (2004) Springer Lexikon Medicine. Springer, Heidelberg

Streckert HJ (2009a) *Astrovirus*. In: Neumeister B et al (eds) Microbiological diagnostic, 2nd edn. Thieme, Stuttgart

Streckert HJ (2009b) *Calicivirus*. In: Neumeister B et al (eds) Microbiological diagnostic, 2nd edn. Thieme, Stuttgart

Streckert HJ (2009c) *Reovirus*. In: Neumeister B et al (eds) Microbiological diagnostic, 2nd edn. Thieme, Stuttgart

Tschäpe H, Reissbrodt R, Prager R (2009) *Salmonella* spp. In: Neumeister B et al (eds) Microbiological diagnostic, 2nd edn. Thieme, Stuttgart

Vobis M, D'Haese J, Mehlhorn H, Mencke N (2003) Evidence of horizontal transmission of the feline leukemia virus by the cat flea (*Ctenocephalides felis*). Parasitol Res 91:467–470

Vobis M, D'Haese J, Mehlhorn H, Mencke N (2005) Experimental quantification of the feline leukemia virus in the cat flea (*Ctenocephalides felis*) and its feces. Parasitol Res 97:S102–S106

WHO (2004) Burden of diseases (Bulletin). WHO, Geneva, Switzerland

Chapter 10
What Are Zoonotic Diseases?

Heinz Mehlhorn

Abstract Humans consider themselves as "crown of the creation," since they feel to be guided by a breeze of the "divine spirit" that enables some insights into the scene of life and into basics of earth. However, with respect to their relationships on earth, humans are just members of the animal kingdom and thus endangered by the same agents of disease that threaten the life of animals. Since humans as predators are also members of the food chain on earth, they may also become infected with agents of diseases that are on or inside of animals that belong to the daily human food. Diseases due to such animal-based or animal-transmitted pathogens are called *zoonosis*. Such pathogens may belong to the groups of prions, viruses, fungi, bacteria, and animal parasites, which may interact in a broad spectrum of pathways. The main topics of this book—*Blastocystis* species—belong to these pathogens. Thus many pathways of transmission of the numerous above-cited pathogens will run identically or are at least very similar. Therefore it is worthwhile to throw a glimpse onto the, in general, already available transmission pathways of the agents of zoonotic diseases while giving definitions and showing important examples.

10.1 Introduction

Since about 20,000–100,000 years, when larger settlements were founded, the specimens of the *Homo sapiens* started to live in close neighborhood to farmed animals and/or animals (such as rats, mice, dogs, flies, etc.), which seek shelter or food inside or close to human dwellings. This was the chance for several, primarily strictly host-specific agents of diseases or of parasites of animals to begin a phase of coevolution inside animals and humans and thus to enlarge their host spectrum

H. Mehlhorn (✉)
Institute for Parasitology, Heinrich Heine University, Universitätsstr. 1,
Düsseldorf 40225, Germany
e-mail: mehlhorn@uni-duesseldorf.de

H. Mehlhorn et al. (eds.), *Blastocystis: Pathogen or Passenger?*,
Parasitology Research Monographs 4,
DOI 10.1007/978-3-642-32738-4_10,

considerably and as consequence to obtain better chances for a long-term survival on earth. Such processes of adaptations were found in viral, bacterial, fungal, and parasitic infections, which after a slow beginning then occurred more and more often simultaneously in the constantly growing human population. Many of these natural cross-infections surely failed, while others probably ended fatal for those hosts—animals or humans—which were infected as a second host after the agent of disease had been propagated for a long time in its first hosts, which therefore had the chance to become adapted. Thus then these first hosts suffered only from low-graded symptoms of disease or even showed no disturbances after an infection. Therefore there are mostly much heavier symptoms of disease in those cases, where agents of diseases hit a host population, which had not yet been in contact with such aggressors. For example, mankind probably suffered longer from infections with *Plasmodium vivax* than with *P. falciparum*, which demands today a much higher death toll per year than all other malaria types together. Similarly opportunistic parasites such as *Giardia lamblia* introduce practically no symptoms of disease in immunocompetent hosts, while they lead often to severe, slimy diarrheas in naïve hosts (=those without any previous contact to the parasite) or in immunocompromised persons (e.g., in AIDS-patients). Another factor supporting the development of zoonotic diseases was given when mankind started farming of animals close to their housings and enlarged it in our days of mass production of animals in overcrowded stables. This close neighborhood of animals and humans (often at short distances to towns with huge populations) increases the chances for ping-pong transmissions of agents of diseases between animals and humans. This is especially the case, when flies that are attracted by feces, get the chance to become vectors by taking up pathogens from feces of some hosts and transport them to the food of other hosts. The potential importance of this often-neglected transportation system was recently shown in a study lasting 3 years, which proved the occurrence of more than 100 important parasites and bacteria on the surface and in the intestine of different fly species which were caught close to stables of horses, pigs, cattle, chicken, or rabbits respectively close to dog ponds or common human recreation sites (Gestmann et al. 2012; Förster et al. 2012). These examinations showed that severe agents of diseases such as EHEC (Enterohemorrhagic *Escherichia coli* bacteria), *Staphylococcus aureus*, *Campylobacter* sp., etc. and/or several common parasites might be easily transmitted. Thus they may introduce severe outbreaks of diseases among crowded populations of humans and animals.

10.2 Which Are the Definitions of Zoonosis?

This term has its origin in the *Greek* words *zoon* = animal and *nosos* = disease that originally meant "disease of animals." The present use of many *Greek* terms as *termini technici* for diseases is based on the facts that this early human high culture left many written documents on diseases and potential medications and that their physicians also used the medicinal knowledge of previous human high cultures such as the Assyrians, Babylonians, Sumerers, or Egyptians. Table 10.1 summarizes some important landmarks in the discovery of zoonotic agents of disease from the early beginning to the early twenties of the last century that was full of discoveries, that are still today of high importance (Mehlhorn 2012a, b, c; Grüntzig and Mehlhorn 2010a, b).

In the more recent past the original sense of the word "zoonosis" as "disease of animals" was erroneously changed by scientists who apparently were not familiar with the "old Greek" language. They started to use the word "zoonosis" to describe a peculiar disease, the agents of which were constantly transmitted in a cycle between animals and humans. However, since there are significant variations in the ways, modes, and directions of the transmission pathways, many international groups of researchers described several subgroups of zoonosis. However, there is no final agreement in the use of the terms among the international community of scientists. In general up to five different types of zoonosis were used in literature (see Mehlhorn 2008, 2012a, b, c).

1. *Direct zoonosis*: This term is often used to characterize the transmission of agents of disease, the shape of which remains identical in animals and later in infected humans. Examples of such direct zoonosis occur during the transmission of the viruses of the rabies disease (*Lyssa*-virus), of the bacteria of the brucellosis (*Brucella* species), of the parasitic protozoans introducing the entamoebiasis (*Entamoeba histolytica*) or the giardiasis (*Giardia duodenalis*). Furthermore the top targets of this book—the *Blastocystis* species—would also fit into the category of a direct zoonosis, since the fecally excreted cyst stages are taken up orally and then may colonize the intestinal tract of humans.
2. *Cyclic zoonosis*: In these cases a successful transmission from one host to another one may only occur after a morphologic transformation of the agent of disease has occurred in an animal or in a human. This example is the case in the life cycles of tapeworms such as *Echinococcus granulosus* or *Taenia solium*, both of which are found as adult worms in the intestine of the final host (=predators, carnivores) and as larval stages in the muscles and/or other tissues of intermediate hosts (prey animals, herbivores or omnivores).
3. *Metazoonosis*: This term is used to describe transmission pathways that include arthropods as distributors (=vectors), which in some cases ingest an agent of disease, allow its reproduction in their bodies or not, but transmit these pathogens to another host during a bite or by fecal contamination. Examples for this category would be the transmission of plague bacteria (*Yersinia pestis*) by fleas, the viruses of the Dengue or Yellow fever by *Aedes* mosquitoes, or the mechanical transmission of bacteria and parasites that are attached at the feet, mouthparts of flies, or that are included in the feces of any other insect (e.g., cockroaches).
4. *Saprozoonosis*: This group contains agents of disease that occur in nature on the soil or on degenerating contaminated food of animals which thus become infected. These pathogens may be transmitted, when humans eat raw meat or come into contact with fecally contaminated plants or waste waters. Examples

Table 10.1 Historical landmarks in the discovery of zoonotic parasitic diseases (selection)

Date	Event
5000 BC	Nematode eggs discovered recently in a frozen human body (Ötzi) in Austrian Alps
2000 BC	*Taenia* and *Schistosoma* ova in Egyptian mummies
1500 BC	Papyrus Ebers (Egypt) gives reference to roundworms (*Ascaris lumbricoides*), threadworms (*Enterobius vermicularis*), and tapeworms (*Taenia* sp.)
1234–1300 BC	Biblical references to *Dracunculus medinensis* in Red Sea region
700–600 BC	Records of *Dracunculus medinensis* worms from Mesopotamia (today Irak)
500 BC	The Greek poet Herodot reported that the Egyptians used fine nets against mosquitoes
430 BC	The Greek physician Hippocrates (460–399 BC) described *Ascaris*, *Oxyuris*, adult *Taenia*, and Malaria
342 BC	The Greek natural scientist Aristotle (384–322 BC) established a first classification system for animals (*Historia animalium*) and described flat and round worms
300 BC	Chinese description of threadworms, tapeworms, hookworms, and hookworm disease
20 AD	The Roman Aulus Celsus recognized as tapeworms *Taenia*, *Tinea*, and *Taeniola* (*Taenia* sp.), vermes cucurbitini (tapeworm proglottids), "hailstones" (cysticerci), and roundworms, lumbrici teretes (*Ascaris lumbricoides*)
62 AD	In his *Historia naturalis*, the Romans Lucius Columella (40–88 AD) and Plinius Secundus (23–79 AD) reported on parasitic animal diseases
129–199 AD	The Greek-Roman physician Galen recognized three types of worms: roundworms (*Ascaris lumbricoides*), threadworms (*Enterobius vermicularis*), and tapeworms (*Taenia* sp.) and also cysticerci in livers of slaughtered animals
1000 AD	Avicenna (Abu Ali El-Hosein Abdallah Ibn Sina, 980–1037). This Persian theologian and physician reported in his book *Liber canonis medicinae* on malaria and many worms, especially on *Dracunculus*, which today in French is still called *Fil d' Avicenne*
1150 AD	Hildegard of Bingen (1098–1179). This German nun published plant based methods of treating worms in her book *De causis et curis morborum*
1498 AD	The Italian monk Girolamo Savonarola (1452–1498) described in his *Tractatus de vermibus* the occurrence and treatment (by mercury) of worm-infected humans
1520 AD	The German physician P.A.T. Bombastus von Hohenheim, called Paracelsus (1496–1541) introduced inorganic salts (e.g., zinc salts) as anthelminthica
1684–1698 AD	The Italian physician and philosopher Francesco Redi (1626–1697) described in his book *Osservazioni interno agli animali viventi* about 108 different worms and published (in 1798) a detailed study on *Fasciola hepatica*. Due to his leadership he is considered the Father of Parasitology—the name "rediae" for special trematode developmental stages honors this
1699 AD	The Dutchman Nicolaas Hartsoeker (1656–1725) and the Frenchman J. Andry (1658–1742) proposed that helminth infections derive from oral intake of excreted worm eggs
1717 AD	The Italian Lancisco postulated that malaria is caused by bites from mosquitoes
1739–1778 AD	The Swedish physician Carl von Linné (1707–1778) created within the different editions of his book *Systema naturae* a comprehensive system of classification of animals and plants
1756 AD	The English physician Alexander Russel (1715–1768) discovered in Aleppo (today in Syria) the skin leishmaniasis

(continued)

Table 10.1 (continued)

Date	Event
1801 AD	The German physician and natural scientist Carl Asmund Rudolphi (1771–1832) published his book *Entozoorum historia naturalis* with the taxonomy of all available parasites
1801 AD	The French natural scientist Jean Baptiste Lamarck (1744–1829) presented the first general theory of the evolution within his *Philosophie Zoologique*
1820 AD	The French scientists Pierre Joseph Pelletier and Joseph Lavandou isolated quinine from plants
1838 AD	The German scientists N. Schleiden and Theodor Schwann (1810–1882) formulated the "cell theory," which became the dead end for the former de novo-creation-theory
1848 AD	The American Josiah Nott postulated again that mosquitoes were vectors of malaria and yellow fever
1850 AD	The German physician Theodor Bilharz discovered *Schistosoma haematobium* in Cairo, Egypt
1853 AD	The German C.T.E. von Siebold (1804–1885) showed the life cycle of *Echinococcus granulosus*
1855 AD	The German physician Gottlieb Küchenmeister demonstrated that *Taenia solium* infections are acquired by eating infected pork
1859 AD	The Germans Rudolph Virchow (1821–1902) and Rudolph Leuckart (1822–1898) independently discovered the life cycle of *Trichinella spiralis*
1859 AD	The English philosopher, theologian and natural scientist Charles Darwin (1809–1882) published his book *The Origin of Species*—another landmark of the evolution theory
1863 AD	The English scientist Thomas Cobbold suggested that snails might be the intermediate host of schistosomes
1867 AD	The German zoologist Rudolf Leuckart described the life cycle of *Echinococcus granulosus*
1869 AD	The German physician Otto Wucherer (1820–1873) discovered in Brazil microfilariae and schistosomal eggs in human urine
1876 AD	The English scientist Joseph Bancroft observed and described adult *Wuchereria banchrofti* worms
1878 AD	The Scottish scientist Sir Patrick Manson (1844–1922) working in China showed that *Wuchereria bancrofti* is transmitted by mosquitoes (*Culex* sp.)
1880 AD	The French Charles L.A. Laveran (1845–1922) showed malaria stages within erythrocytes (Nobel Prize in 1907)
1881 AD	Leuckart and Thomas independently described the life cycle of *Fasciola hepatica*
1888 AD	The French physiologist Charles Richet (1855–1935) formulated the basic concept of humoral immunity (Nobel Prize 1919)
1893 AD	The Americans Theobald Smith and F.L. Kilbourne identified the transmission of *Babesia bigemina* by ticks (*Boophilus annulatus*)
1895 AD	The English military physician David Bruce (1855–1932) showed that the tsetse fly is the vector of animal trypanosomes
1897 AD	In India the English Army doctor Sir Donald Ross (1857–1932) proved that avian malaria was transmitted by *Anopheles* mosquitoes (Nobel Prize in 1902). In the same year the Italians Bignami, Bastianelli and Grassi did the same for human malaria

(continued)

Table 10.1 (continued)

Date	Event
1898 AD	The German physician and discoverer of the agents of anthrax, cholera and tuberculosis (*Mycobacterium tuberculosis*) Robert Koch (1843–1910) described *Theileria parva*, the agent of East Coast fever (Nobel Prize 1905)
1898 AD	The French P.L. Simond succeeded in demonstrating the transmission of plague by rat fleas
1900 AD	In Cuba an American group working with Walter Reed demonstrated the transmission of yellow fever by mosquitoes (*Aedes aegypti*)
1903 AD	The English scientists W.B. Leishman and C. Donovan independently described *Leishmania donovani*, the agent of Kala-azar disease (Leishmaniasis)
1904 AD	In Cairo the German helminthologist Arthur Loos (1861–1923) discovered the transmission of the hookworm
1906 AD	The American physician Howard T. Ricketts (1878–1910) recorded the tick *Dermacentor andersoni* as being a vector of the agents of the Rocky Mountain spotted fever
1906 AD	The German zoologist and discoverer of the agent of syphilis Fritz Schaudinn (1871–1906) described *Entamoeba histolytica* as a human parasite introducing bloody diarrhea
1907 AD	The American E.E. Tyzzer described stages of the genus *Cryptosporidium*
1907 AD	The German chemist Paul Ehrlich (1854–1925)—the father of chemotherapy—proposed the drug trypan red against trypanosomiasis (Nobel Prize 1908)
1908 AD	In North Africa the French military doctors Charles Nicolle (1866–1936) and L.H. Manceaux described *Toxoplasma gondii* in a rodent
1909 AD	The Brazilian Carlos Chagas (1979–1934) discovered the life cycle of *Trypanosoma cruzi* and described *Pneumocystis carinii*
1909 AD	Teams working with the French scientist Charles Nicolle (1866–1936) (Tunis) and the American H.T. Ricketts (Mexico) proved that the louse *Pediculus humanus corporis* is the vector of the typhus-causing rickettsia
1910 AD	The Italian scientist Carini discovered *Pneumocystis carinii* in rats
Until today	And so on the work of hundred thousands scientists, etc

for this group would be the transmission of the bacterial agents of listeriosis, of the spores of fungi of mycosis, or cysts of amoebic dysenteria.

5. *Other authors* grouped zoonotic transmission procedures according to the main direction that is taken by an agent of disease.

    (a) *Zooanthroponosis*: In these cases the infectious agents occur mainly in animals and may be transmitted from there to humans. Examples for this category would be the transmission of the *Trichinella spiralis* larvae in muscles of wild boars to humans when eating undercooked meat. Also the transmission occurs when *Balantidium coli* cysts are swallowed by humans.

    (b) *Anthropozoonosis*: This term describes the pathway of transmission, when a pathogen of an animal has its origin in humans and leads to disease in animals. This may be the case when e.g., wild bears or wolves feed on humans infected with *Trichinella* muscle larvae or when pigs or cattle swallow human feces containing the sporocysts and/or the oocysts of *Sarcocystis* species. Of course pathways of transmission starting from humans are much more rare than those with a source in animals reaching humans from there.

When looking at the great variations of transmission pathways that have been established by zoonotic agents between humans and animals during coevolution, an absolutely correct classification according to the categories described above is mostly very difficult. Therefore the proposal to use exclusively the term *zoonosis* to characterize any of the numerous ping-pong infections between humans and animals seems most reasonable. In this case, however, it makes sense to differentiate according to the main pathways of transmission into:

(a) Food-borne zoonotic diseases
(b) Vector-borne zoonosis
(c) Cyclic zoonosis
(d) Contamination-borne zoonosis

10.3 Examples of Zoonotic Pathways in the Transmission of Pathogens Between Humans and the Animals and from Their Surroundings

Except for prions and viruses which depend on living cells and are reproduced by them all other members of the living society on earth—plants, fungi, animals (including humans)—have to fight for their survival, i.e., they have to defend their living space (=biotope) against competitors. Only those, which had been the winners in this daily survival fight, are our contemporaries. All members of the living society on earth belong to a giant food chain—a pyramid with predators at top sites and preys and producers at the lower levels. Therefore it is not astonishing that there had been established correlations among the different members of the food chain and that man and animals are attacked by various organisms seeking shelter and propagation inside their bodies. Such invasions may occur occasionally or might have been established since long and thus run often according to extremely fixed pathways of transmission. Tables 10.2, 10.3, 10.4, 10.5, 10.6 in the following Sects. 10.3.1–10.3.5 list some of the most important human diseases, which have their sources in organisms that exist longer on earth and thus in the human surroundings of our days than mankind. This is not only true for the rather recent *Homo sapiens* with his only ~100,000-years-old history but also for its precursors such as specimens of the genera *Ardepithecus* or *Australopithecus*, etc. Since man still exists although there are these huge amounts of hostile agents of diseases around him, there is good hope that humans will have a future on earth.

Table 10.2 Prions with a zoonotic potential

Species	Characteristics	Disease/treatment	Symptoms of disease	Hosts	Pathway of transmission	Geographic distribution
Prions of the variation of Creutzfeld-Jacob (Bovine Spongious Encephalopathy; BSE)	Infectious proteins, misfolded in β-constellation: PrP^{sc}	New variation of Creutzfeld-Jacob/no treatment	Grimasses, giddiness, ataxia, brain destruction, death	Humans, cattle	Eating infected beef, blood transfusion, iatrogen (during surgery)	Worldwide; epidemic in Great Britain (1995–1998), a few cases in France, USA, Spain, Italy, Canada, etc.
Prions of Kuru	Infectious proteins, misfolded in β-constellation: PrP^{sc}	Kuru/local name: laughing death/ no treatment	Grimasses, tremor, brain destruction, death; local name: laughing death	Humans	Cannibalism	Papua-New Guinea
Scrapie-prion	Infectious proteins, misfolded in β-constellation: PrP^{sc}	Scrapie in sheep/ (German: Traberkrankheit)/ no treatment	Sticksy walking of sheep and cattle, turning movement	Sheep, cattle	Cows and sheep are probably infected by eating fly larvae in grass that had previously fed on infected sheep bodies or on placental material (our group showed experimental transmission via fly larvae to hamsters)[a]	Worldwide

[a]Post et al. (2009)

Table 10.3 Virus-borne diseases with zoonotic potential (selection)

Family/genus	Virus	Main symptoms of disease	Hosts	Pathway of transmission	Geographic dislocation
Togaviridae *Alphavirus*	Sindbis	Fever, exanthema	*Humans*, birds	Bite of mosquitoes	Europe, Africa, Asia, Australia
	Chikungunya	Fever, pain in joints	*Humans*, monkeys	Bite of *Aedes*-mosquitos	Africa, Southern Europe
	American horsevirus	Fever, encephalitis	*Humans*, horses	Bite of midges, mosquitoes	North-, South America
Flaviridae *Flavivirus*	Dengue fever	Hemorrhagic fever, classic fever	*Humans*, monkeys	Bite of mosquitoes (*Aedes*)	All tropics and subtropics
	Japanese Encephalitis	Encephalitis, flu, mortalitiy ~25 %	*Humans*, birds	Bite of mosquitoes (*Aedes*)	North, South Asia, India
	West-Nile fever	Fever, encephalitis	*Humans*, birds, dogs, horses, etc.	Bite of mosquitoes (*Aedes*, *Culex*)	Africa, South America, Israel, South France
	Yellow fever	Fever, icterus, mortality up to 50 %	*Humans*, monkeys	Bite of mosquitoes (*Aedes aegypti*)	South America, Africa
	Spring-summer-meningo-encephalitis	Meningitis, encephalitis, fever, mortality ~10 %	*Humans*, mice, many other animals	Bite of *Ixodes* ticks	Russia, Central Europe, countries of the Balcony
Bunyaviridae	Hanta	Lung inflammations, hemorrhagic fever	*Humans*, mice, rodents	Fecal-orally, flies	Europe
	Phlebo	Fever, meningitis	*Humans*	Sandfly bite	Mediterranean Sea
	Rift-Valley	Fever, encephalitis, hepatitis	*Humans*, cattle, sheep, ruminants	Bite of midges, many mosquitoes	Central, South Africa, Egypt
	Crimean-Congo	Haemorrhagic fever	Cattle, *humans*	Bite of *Hyalomma*-ticks	Africa, Turkey, Black sea region

(continued)

Table 10.3 (continued)

Family/genus	Virus	Main symptoms of disease	Hosts	Pathway of transmission	Geographic dislocation
Reoviridae Coltiviren	Colorado-tick-virus	Encephalitis	*Humans*, roe deers	Bite of ticks	Rocky-Mountains, USA
Seadorna	Banna-virus	Encephalitis	*Humans*	Fecally, fly contact	China, Indonesia
Rotaviruses	*Rotavirus* (A-G)	Gastroenteritis, diarrheas, enterotoxicity	*Humans*, many animals	Oral-fecal,fly contacts	Worldwide
Rhabdoviridae *Lyssavirus*	Lyssa (rabies)	100 % death toll	*Humans*, foxes, dogs, cats, wolves, roe deers	Bite of animals, smear infection, wound contamination, flies?	Worldwide
Orthomyxoviridae	Influenza-virus	High fevers, lung infection	*Humans*, animals	Aspiration, fly transmission	Worldwide
Calciviridae	Calici-virus	Diarrheas	*Humans*, animals	Oral-fecally,flea feces and bites	Worldwide

Table 10.4 Bacteria of zoonotic importance (selection)

Species	Characteristics	Disease/treatment	Symptoms of disease	Hosts	Pathway of transmission	Geographic distribution
Anaplasma phagocytophylum (syn. *Ehrlichia*)	Coccoid, gram-negative, pleomorphe, nonmotile, obligatorily intracellular bacteria of 0.4–1.9 μm in size	Human granulocytic anaplasmosis (HGA)/ Doxycyclin	High fevers, occasional meningitis, exanthemes	*Humans*, ruminants, roe deers, dogs, cats, mice	Bite of *Ixodes* ticks	North and Central Europe, North and East of USA
Bartonella (syn. *Rochalimaea*) *quintana*	Microaerophilic, Gram-negative, facultatively intracellular rods	French fever; 5 days fever; bacillary pelliosis; Wolhynic fever/ Doxycyclin/Rifampicin	Bacteremia, endocarditis	*Humans*	Inhaling of feces of body lice	Worldwide
Bartonella bacilliformis	Microaerophilic, Gram-negative, facultatively intracellular rods	Carrion disease, Oroya-fever, Verruga peruana/ Doxycyclin	High fevers, intraerythrocytic bacteriamia	*Humans*	Bite of sandflies	Peru, Ecuador
Bartonella henselae	Microaerophilic, Gram-negative, facultatively intracellular rods	Cat-scratch-disease/ Azithromycin in severe cases; normally not needed	Lymphadenitis	*Humans*, cats	Scratching by cats or by contaminated mouthparts of fleas and ticks	Worldwide
Borrelia burgdorferi, *B. afzelii*, *B. garinii*, etc.	Spirillic with irregular windings 0.2–0.5 × 8–30 μm, possess endoflagella	Borreliosis; Lyme borreliosis/Doxycyclin	Rosacea migrans, Acrodermatitis; arthritis, paralysis	*Humans*, many warm-blooded animals	Tick bites (*Ixodes* species)	Europe, USA
Borrelia duttoni	Spirillic with irregular windings 0.2–0.5 × 8–30 μm, possess endoflagella	Tick-borne relapse fever/ Cephalosporines	High fever, myalgia, icterus, hepatomegaly, high lethality	*Humans*, reservoir mice, camels	Bites of *Ornithodorus moubata*-ticks	Africa
Borrelia recurrentis	Spirillic with irregular windings 0.2–0.5 × 8–30 μm, possesses endoflagella	Louse-borne relapse fever/ Cephalosporines	High fever, myalgia, icterus, hepatomegaly, high lethality	*Humans*	Inhaling of contaminated body louse feces	Worldwide
Borrelia turicata, *B. parkeri*, etc.	Spirillic with irregular windings 0.2–0.5 × 8–30 μm, possesses endoflagella	American tick-borne relapse fever/Doxycyclin	High fever, myalgia, icterus, etc. hepatomegaly, lethality	*Humans*, mice, rodents	Bites of *Ornithodorus* species (argasid ticks)	Northwestern America

(continued)

Table 10.4 (continued)

Species	Characteristics	Disease/treatment	Symptoms of disease	Hosts	Pathway of transmission	Geographic distribution
Brucella melitensis, *B. suis*, *B. canis*	Gram-negative, coccoid rods; reproduction in macrophages	Brucellosis, Malta-fever; Mediterranean fever; Bang disease/ Doxycyclin + Rifampicin	Severe influenca-like; lethality 2–5 %; abortions	*Humans*, sheep, goats, cows, dogs, pigs	Infected raw meat, raw milk, fly contact?	Worldwide in endemic regions
Campylobacter species; *C. jejuni*, *C. coli*	Spirillic bacteria (with flagella) of 0.5–5 μm in length inside intestinal lumen	Enterocolitis/ Fluorchinolones, Erythromycinstearat	Enterocolitis with fever; Guillain-Barré-Syndrom	*Humans*, birds, cattle, pigs	Fly transmission or oral uptake of bacteria from bird feces, raw meat of milk.	Worldwide
Coxiella burnetii	Gram-negative, obligatorily intracellular bacteria; similar to legionellids	Q-fever (Query-fever)/ Doxycyclin	Like a summer flu	*Humans*, sheep	Bites of about 50 tick species	Worldwide
Ehrlichia chaffeensis, *E. ewingii*	Coccoid, Gram-negative, pleomorphe, nonmotile, obligatorily intracellular bacteria of 0.4–1.9 μm in size	Human monocytic ehrlichiosis/Doxycyclin	High fevers, occasional meningitis, exanthemes	Dogs, wild ruminants, *humans*	Bites of *Amblyomma*-ticks	Southern USA
Francisella tularensis, *F. holarctica* and other subspecies	Gram-negative rods with capsule; intracellular reproduction	Tularaemia/Gentamycine, Streptomycine	High fevers, respiratory and/or dermal syndromes; lethalitiy up to 50 %	*Humans*, rodents, hares, wild animals	Inhalation, contaminated mouthparts of arthropods	Northern hemisphere, East Asia
Helicobacter species; *H. pylori*, *H. suis*, *H. heilmanni*, etc.	Spirillic bacteria (with flagella) of 3–6 μm in length; inside stomach (closely attached to antrum and corpus regions of stomach)	Gastric and hepato-intestinal symptoms/ Clarithromycine, Amoxicillin, etc.	Chronic gastritis, often symptomless	*Humans*, monkeys, pigs, etc.	Fly transmission or oral uptake of fecally contaminated food	Worldwide
Legionella pneumophila	Coccoids, reproduction in macrophages and in water amoebae	Legionnaire's disease; Pontiac fever/ Macrolids, Rifampicin	Pneumomonia; lethality up to 80 %	*Humans*, free water amobae	Inhalation of infected amoebae in biofilms of water systems	Worldwide

Leptospira species; *L. interrogans*; *L. borgpetersenii*, etc.	Spirillic, very thin, highly motile bacteria 0.2 × 6–30 μm in size	Weil's disease, leptospirosis/Penicillin G, Doxycyclin	Septicaemia, fever, diarrhea, icterus	*Humans*, wild animals, rats, mice	Fecal contamination of wounds, inhaling bacteria from feces of rats	Worldwide
Listeria monocytogenes (syn. *Corynebacterium*), *infantisepticum*	Facultative anaerobe rod-like Gram-positive, peritriche bacteria, intra- and extracellular	Listeriosis/Ampillicin	Flu-like symptoms, but also meningitis, evtl. abortus	*Humans*, ruminants, pigs, birds	Fecal contamination, milk of animals	Worldwide
Neorickettsia sennetsu (syn. *Ehrlichia*)	Coccoid, Gram-negative, pleomorphe, nonmotile, obligatorily intracellular bacteria of 0.4–1.9 μm in size	Japanese sennetsu-ehrlichiosis, mononucleosis/ Doxycyclin	Fever, hepatosplenomegaly, leucopenia	*Humans*	Tick bites	Japan, Malaysia
Orientia (syn. *Rickettsia*) *tsutsugamushi*	Obligatory intracellular short or coccoid rods of 0.8–2 μm in size	*Tsutsugamushi*-fever/ Doxycyclin	High fevers, skin exanthemes	*Humans*, dogs, rodents	Bites of mites (*Trombicula* sp.)	Asia, India, Australia
Rickettsia conorii	Obligatory intracellular, short or coccoid rods of 0.8–2 μm in size	Mediterranean tick-bite fever/Doxycyclin	High fevers, skin exanthemes	*Humans*, dogs, rodents	Bites of *Rhipicephalus sanguineus*-ticks	Mediterranean Sea region, India, Africa
Rickettsia prowazekii	Obligatory intracellular short or coccoid rods of 0.8–2 μm in size	Louse-borne spotted typhus, Brill-Zinsser disease up to 60 % lethality/ Doxycyclin	High fevers, skin exanthemes	*Humans*	Bites and feces of body lice	Worldwide
Rickettsia rickettsii	Obligatory intracellular, short or coccoid rods of 0.8–2 μm in size	Rocky-Mountains spotted fever/Doxycyclin	High fevers, skin exanthemes	*Humans*, dogs, rodents	Bites of *Dermacentor*, *Rhipicephalus*, *Amblyomma* ticks	North and South America
Rickettsia typhi	Obligatory intracellular short or coccoid rods of 0.8–2 μm in size	Murine spotted fever/ Doxycyclin	High fevers, skin exanthemes	*Humans*, rodents	Bites of fleas, e.g., *Xenopsylla cheopis*	Worldwide

Table 10.5 Fungi with an anthropophilic zoonotic potential (selection)

Species	Characteristics	Disease/treatment	Symptoms of disease	Hosts	Pathway of transmission[a]	Geographic distribution
Pneumocystis carinii[b], *P. jiroveci*	Thin-walled cysts with 5–7 μm in diameter, containing up to eight 1 μm sized trophozoites	Pneumocystosis in immunodeficient persons/ Sulfonamids	Lung exsudate, oedema, pneumonia, secondary bacterial super infections	*Humans*, rats	Inhalation of cysts or by sputum contact; fly transmission?	Worldwide
Microsporum canis (syn. *M. equinum*, *M. distortum*)	Colonies grow quickly, borders irregular; microconidia, macroconidia with enlarged tip	*Tinea capitis*, *T. corporis*/ Terbinafin, Fluconazol, Griseofulvin	*Tinea capitis*, *T. corporis*	*Humans*, cats, dogs	Body contact with infected pet animals	Worldwide
Trichophyton rubrum	Slowly growing colonies are flat and appear yellow reddish-brown; mostly no macronoids	All human types of tinea (*T. capitis*, *T. unguinum*)/ Ciclopirox or Amorolfin-nail-set	All human forms of tinea	*Humans*, animals as transporters	Body contact, common use of shoes, clothes	Worldwide
Trichophyton interdigitale	Quickly growing colonies with a witish surface and yellow borders, microconids: coccoid shaped; macroconids 7–50 μm in diameters	*Tinea corporis*, *T. pedis*/ Terbinafin	All human forms of tinea	*Humans*, cats, mice	Body contact, common use of shoes, clothes	Worldwide
Trichophytum equinum	Quickly growing whitish colonies, later brown	*Tinea corporis*/Terbinafin, Fluconazol	All human forms of tinea	*Humans*, horses	Contact with contaminated hair of horses	Worldwide

Trichophyton benhamiae complex	Like *T. interdigitale*	*Tinea capitis*, *T. corporis*/ *T. capitis*: Griseofulvin, Fluconazol; *T. corporis*: Terbinafin, Fluconazol	Infestations of skin	*Humans*, rabbits, guinea pigs	Contact with hair of infected animals	Worldwide
Candida albicans, and many synonyms	Whitish colonies, pseudomycel, mycel, chlamyclospores	Candidosis, e.g., vaginitis/ Nystatin, Fluconazol, Itraconazol	Infestations of skin and inner mucuous layers, soor, fluor	*Humans*	Body contact, contact to fecal smears, transmission by licking flies	Worldwide

[a]All permanent stages of fungi might be transmitted via contaminated mouthparts or feet of flies

[b]The systematic position of the genus is still unclear—even after molecular biological investigations—since there are characteristics of fungi as well as those of animal cells of protozoan type. Mostly there is no differentiation done between *P. jiroveci* (first described in humans) and *P. carinii* (first described in rats)

Table 10.6 Parasites with a zoonotic life cycle (selection)

Species	Characteristics	Disease/treatment	Symptoms of disease	Hosts	Pathway of transmission	Geographic distribution
Protozoa						
Giardia duodenalis-assemblages	Trophozoites with 8 flagella, cysts with 4 nuclei, 10–20 μm in size	Giardiasis/metronidazol	Slimy diarrheas	*Humans*, many animals	Oral uptake of cysts from feces; fly transmission	Worldwide
Cryptosporidium hominis and animal assemblages	Oocysts are tiny, spherical, measure only 5 μm and contain no sporocysts, but only 4 sporozoites	Cryptosporidiosis/no definite treatment available	Extreme watery diarrheas in immune-suppressive persons, none in immunocompetent ones	*Humans*, many animals	Oral uptake of oocysts from feces; fly transmission	Worldwide
Toxoplasma gondii	Spherical oocysts (10 μm) in cat feces; trophozoites (7–8 μm) in macrophages and tissue cysts in muscle fibers and brain cells of warm blooded hosts	Toxoplasmosis/sulfonamides	None in immuno competent persons; brain damages in immunocompromised ones; fetal damages possible, if pregnant woman had not been infected before	*Humans*, many animals	Oral uptake of oocysts from cat feces Eating of raw or undercooked meat of infected animals Fly transmission	Worldwide
Plasmodium species	Schizogony in liver cells Schizogony in red blood cells Gamogony and sporogony in mosquitoes	Malaria/mefloquin, proguanil, atovaquone, arteminisine, etc.	High, often rhythmic fevers, high mortality in cerebral malaria	*Humans*, some monkeys	Bite of 60 *Anopheles* mosquito species	Tropical countries around the world
Sarcocystis suihominis, *S. bovihominis*	Humans excrete oocysts/sporocysts, in pigs (or cattle) occur schizonts, muscle and brain cysts	Sarcosporidiosis	Humans: Watery diarrheas of high intensity 4–24 h after eating raw meat	*Humans*, pigs	Humans eat raw or undercooked meat of pigs; pigs feed human feces with oocysts; fly transmission	Worldwide
Microsporidia						
Encephalitozoon cuniculi	Spores with an extrudable hollow tube, through which the plasm enters host cells	Encephalitozoonosis/albendazole	Destruction of kidney and other cells	Rabbits, dogs, immune suppressed *humans*	Humans: oral uptake of spores from rabbit feces or other infected hosts	Worldwide

Flukes						
Schistosoma species	♀ and ♂ live as couples inside of blood vessels, eggs possess characteristic hooks at the surface of the shell	Schistosomiasis/ praziquantel	Blood in feces or urine (depending on the species); as late symptoms development of cancer in liver or bladder	Finals hosts: *humans* and many mammals; intermediate hosts: water snails	Skin penetration of cercariae that had been excreted by hosts via urine or feces into waters	Tropical countries
Clonorchis- and *Opisthorchis* species	Flukes are hermaphroditic animals, tiny eggs possess an operculum = cover	Clonorchiasis/ opisthorchiasis/ praziquantel	Liver problems, cancer development	Finals hosts: *humans*, cats, dogs; intermediate hosts: snails and fish	Eating raw or undercooked fish meat with metacercariae	Asia
Tapeworms						
Taenia solium	Adult worms show 4 suckers and a Rostellum with hooks; eggs = embryophores contain the oncosphaera larva	Cysticercosis, taeniasis/ praziquantel	Adult worm: Disturbance of intestinal process;Larva: severe brain problems in case of brain cysticercosis	*Humans*, pigs	Humans: eating raw or undercooked pork meat;pigs: oral uptake of worm eggs in human feces	Worldwide
Dipylidium caninum	Excreted terminal proglottids appear like rice-grains	Dipylidiasis/praziquantel	Intestinal disturbances	Dogs, cats, *humans*	Engorging cysticercoid larvae in portions of cracked fleas inside hair of animals	Worldwide
Diphyllobothrium latum	Up to 20 m long, terminal proglottids are empty	Diphyllobothriasis/ praziquantel	Intestinal problems, anemia of the perniosa type	*Humans*	Eating raw fish with plerocercoids	Worldwide
Nematodes						
Trichinella spiralis	Larva parasitizes intracellularly in muscle cells	Trichinosis/bendazoles	Breathing problems due to intoxication and malfunction of muscles	*Humans* and many meet feeding animals	Eating raw or undercooked meat of infected animals	Worldwide
Anisakis species	Only larvae are found in human intestine	Anisakiasis/bendazoles macrolids	Intestinal problems up to penetration of the intestinal wall; peritonitis	*Humans*, fish	Eating raw solt water fish	Worldwide
Wuchereria bancrofti	Microfilariae free in blood, macrofilariae have their sites in lymph nodes	Filariasis/ diethylcarbamazin/ doxycyclin	Tropical elephantiasis = swelling of legs, breasts, etc.	*Humans*, monkeys	Bite of typical mosquitoes (*Aedes*, *Anopheles*, *Culex*)	Tropical countries

(continued)

Table 10.6 (continued)

Species	Characteristics	Disease/treatment	Symptoms of disease	Hosts	Pathway of transmission	Geographic distribution
Ochocerca volvulus	female worms lay as bundles in skin nodules, larvae are found in lymph below skin	River blindness, onchocerciasis/ ivermectin, doxycyclin	Itching, skin pealing, blindness due to degeneration of larvae inside eyes (as late symptom)	*Humans*, monkeys	Bite of simuliids (*Simulium* species)	Africa, South and Middle America
Acanthocephala						
Macracanthorhynchus hirudinaceus	Intestine-less worms with a hooked anterior pole	Acanthocephaliasis/ loperamid	Intestinal wall perforations, intestinal malfunctions	Pigs, *humans*	Eating of beetles or parts of them containing larvae	Europe, Asia
Pentastomids						
Linguatula serricata	Typical anterior pole with rows of hooks	Pentastomiasis/ albendazoles	Severe liver problems	Dogs, *humans*	Eating undercooked, larvae-containing meat	Europe, Asia

Fig. 10.1 Macrophoto of a pupa and an adult of the gray flesh fly *Sarcophaga carnaria*

10.3.1 Prions with a Zoonotic Potential

Prions—an abbreviation of the description as proteinaceous infectious particles—are no true living organisms but wrongly folded intracellular proteins inducing other proteins to do the same under peculiar not yet well-understood conditions (Prusiner 1998; Rabenau 2009). These wrongly folded proteins, which e.g., in the case of the BSE-epidemiology (Bovine Spongious Encephalopathy) in Great Britain asked a death toll of more than 200 persons and left hundres other victims with severe brain damages besides a loss of more than 500,000 cattle, which had been killed either due to acute infections or due to preventive measurements. Since there was a clear proof for transmission from cattle to humans these diseases were also called transmissible encephalopathies. Similar degenerative symptoms may also occur due to hereditarily transmitted destructions in the genome of some families, but are rare compared to the wave of infections that rolled on after the feeding of animal meet powder (produced from BSE-infected animals) to uninfected ones. The reason was that no one had expected that such powdered animal meat should have been heated up to at least 141 °C in order to destroy these aggressive wrongly folded proteins, while normal cell proteins degenerate mostly already at 60 °C.

In nature there exists a broad potential of transmission of such infectious proteins from one host to the other. This was shown by the experiments of our group (Post et al. 1999). During these experiments *Sarcophaga carnaria* fly larvae were fed with brains of Scrapie-infected hamsters (see Table 10.2). One to seven days later these maggots were squeezed and the minced material was orally inoculated to uninfected hamsters. The same was done with pupae and adult flies (Fig. 10.1) obtained from the originally infected fly larvae. In all cases the hamsters showed 3–5 months after the experimental infection the full range of symptoms of BSE respectively Scrapie-disease including the full loss of motility control. Furthermore they were all prion-positive as was shown by molecular biological tests.

These experiments strongly supported the supposition that apparently these wrongly folded proteins are parts of nature being transmitted since long (probably without producing broad outbreaks but only single fatal cases). Such a transmission scenario would be given e.g., in cases when fly larvae feed on the remnant placenta of an infected animal and were later engorged by other grass feeding ruminants.

Fig. 10.2 Macrophoto of the adult green flesh fly *Lucilia sericata*

Such a transmission will also explain the astonishing findings during the BSE-outbreak in Europe that some cows that had definitively not been fed with animal meat powder were positively tested for BSE.

However, evaluating all these aspects together, it can be stated that prions (Table 10.2) will not be a considerable threat for human health. Under natural conditions (i.e., when ruminants are fed by normal grass or plants) the natural infection rate of those animals with prions will never lead to such a significant increase of the amounts of prions inside the ruminant populations, that this later could lead to a significant infection potential.

10.3.2 Virus-Borne Zoonotic Diseases

The term virus has its origin in the *Latin* word virus, which means mucus, poison, moisture. Viruses themselves are no living organisms but are obligatory based as intracellular pathogens on the reproduction machinery of their host cells, which is situated in the nucleus and along the ribosomes. Therefore successful viruses had developed methods to become firmly attached to potential host cells, to enter such cells, and to find their way to the steering system, where their genetic material—RNA or DNA (in species specifically varying arrangements)—might be reproduced and finally released from the invaded cell. Thus for viruses it is most important to reach the surfaces of such cells otherwise they will burst or become digested by cells of the defense systems of attacked hosts. Therefore the viruses have developed some major pathways of transmission (Table 10.3):

- Transmission by vector bites (Fig. 10.3)
- Transmission by feces of vectors (Figs. 10.1, 10.2, 10.3, 10.9a, b)
- Transmission by feces of infected hosts
- Transmission by contact or by inhaling of excretions or raw meat of infected hosts

While it is much easier for humans to avoid the last three transmission pathways using normal hygienic preventive methods, the contact to vectors occurs practically unlimited—at least in low numbers. Therefore the transmission and propagation of the pathogens of the so-called *emerging diseases* (Mehlhorn 2012a) increases

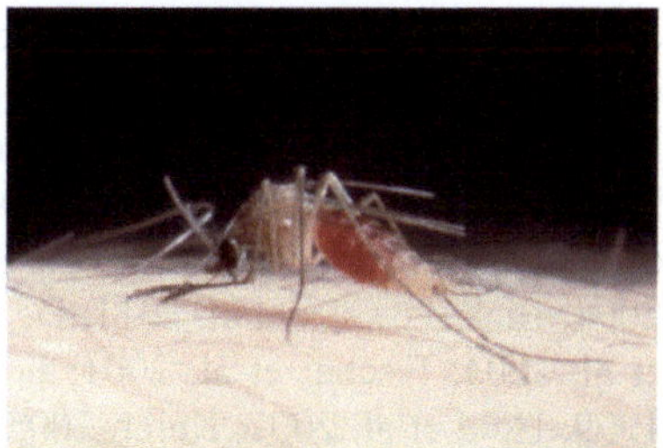

Fig. 10.3 Macrophoto of an *Aedes* mosquito sucking blood on human skin

constantly under the recent worldwide accelerating conditions of globalization and global warming (i.e., local change of the climate and huge migration and traveling activities worldwide; Dobler and Aspöck 2008, 2010; Neumeister et al. 2009).

Table 10.3 summarizes important viral diseases of humans which are discussed in more detail in Sects. 10.4–10.7 (Aspöck 2010; Dobler and Aspöck 2008, 2010; Darai et al. 2009; Löscher and Burchard 2010; Neumeister et al. 2009). The existence of viruses and their high potential to create new combinations, however, endangers life on earth, since they are easily transferred by genetic variations into highly dangerous pathogens. Due to their reproduction inside host cells there are only very restricted possibilities of chemotherapy apart from vaccination trials, which, however, are very often not successful, since many viruses often change their surfaces that have to be recognized by vaccines.

10.3.3 Bacteria of Zoonotic Importance

The present life on earth would not be possible without bacteria. They are found everywhere in nature, inside body cavities, and they even may settle inside cells. Since long some of these bacteria have undergone (as our mitochondria) a close symbiosis with all living organisms. However, a broad spectrum of these organisms that are invisible with naked eyes, may also introduce severe diseases, which in times of worldwide epidemics and pandemics of plaque or cholera reduced mankind significantly in many regions. For example, in Europe the population was reduced during the so-called 30 years-war (1618–1648) down to 15 % mainly by plaque epidemics. Even today the bacteria of tuberculosis or cholera request millions of victims every year.

It is not long ago when both "popes of bacteriology," the German *Robert Koch* (1843–1910) and the French *Louis Pasteur* (1822–1895) and their famous scholars started to enlighten the hidden world of "good and bad" *bacteria* and to develop preventive measurements, which later were supported by the invention of the sulfonamides by the German Bayer Company scientist Gerhard *Domagk* (1895–1964) and the penicillin by the Scottish scientist *Alexander Fleming* (1881–1955). Both got (1939: Domagk and 1945: Fleming) the Nobel Prize for Medicine. However, the bacteria struggled back by a constant and often very quick development of resistances, so that today many bacterial diseases are far from being under control (and even may start a new career as "emerging diseases."

Table 10.4 summarizes a selected spectrum of bacterial diseases, which are spread between humans and animals in their surroundings. The peculiar pathways of transmission are discussed in the Sects. 10.4–10.7 of this article (Alam and Zurek 2004; Bielaszewska et al. 2011; De Jesus et al. 2004; Ekdahl et al. 2005; Emerson et al. 2004; Fischer et al. 2001; Fotedar 2001; Fotedar et al. 1992; Grübel et al. 1999; Hold et al. 2007; Karch 2005; Kobayashi et al. 1999; Mian and Jacal 2002; Nichols 2005; Olsen and Hammack 2000; Rasko et al. 2011; Sasaki et al. 2000; Szalanski et al. 2004). Again the vector-transmitted pathogens of this group are most difficult to control, since the bites of the mostly lonely acting blood suckers (ticks, mites, mosquitoes, tabanids) or contact to skin lickers (flies) can only hardly be avoided.

10.3.4 Fungi with an Anthropophilic Zoonotic Potential

Fungi, the name of which has its origin in the *Latin* word: fungus = mushroom are heterotrophic organisms, which are covered by a chitin containing outer layer consisting of a network of fibrils. Most species of the fungi belong to the free nature, settle and feed at moist places. Several species of the fungi (*French*: champignons) are eatable.

However, some of the fungi species (see Table 10.5) invade body cavities of humans or animals respectively spread on their skin—mostly at places, where a high humidity is guaranteed as it is in the case at the hair-covered scalp, below finger or toe nails, in the groins, or in the shoulder-hollows. These places will be invaded especially in patients with immune suppression and can be treated only with difficulties (Figs. 10.4 and 10.5). This is due to the hidden places of the infection, the poor penetration of chemical substances through the surface of the fungi and due to a general lack of really potent fungicides with low side effects. The main transmission pathways are contacts to contaminated surfaces, direct contacts to contaminated human skin or to hair of infected animals. Other possibilities of transmission of fungi from human or animal feces or from other excretions are given via contaminations of food or materials (Abbot 2002; Banjo et al. 2005; Darai et al. 2009; Neumeister et al. 2009).

10.3.5 Parasites with a Zoonotic Life Cycle

Parasites are per definition in a strict sense animals that live on costs of other animals. Of course there also exist plants that stay as parasites on other plants, but they are here not considered, since they do not harm the health of humans. The name "parasites" has its origin in the *Greek* word "parasitos," which described

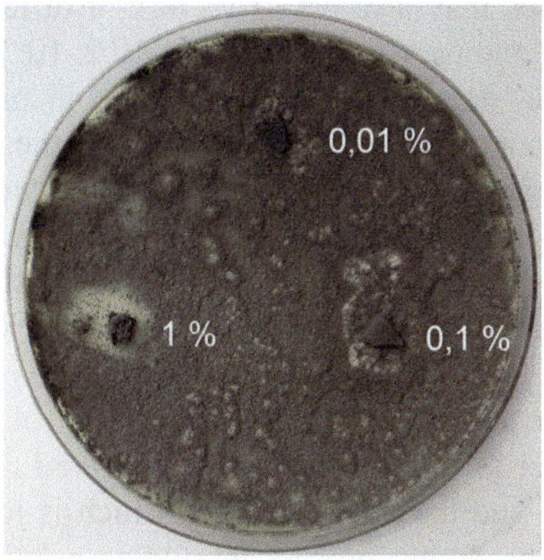

Fig. 10.4 Macrophoto of a culture of *Aspergillus fumigatus* which may enter into the lung of mammals after injection of three different concentrations of the neem-seed derived extract MiteStop (Fa. Alpha-Biocare). Note the clear, doses dependent spots of clearance

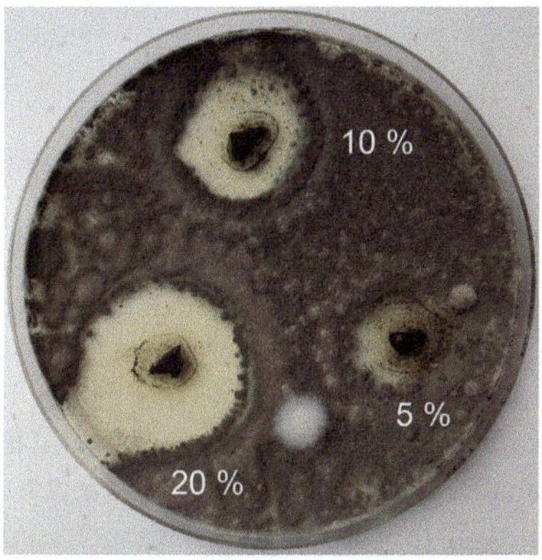

Fig. 10.5 Macrophoto of a culture of *Aspergillus fumigatus* which may enter into the lung of mammals after injection of three different concentrations of the neem-seed derived extract MiteStop (Fa. Alpha-Biocare). Note the clear, doses dependent spots of clearance

"employees" at noble courts that had to taste the food for poisons and thus were nourished on foreign costs. The parasites attack animals and humans. However, although many parasitic species rely on many hosts (e.g., *Toxoplasma gondii*), some had become specialists in the last 10,000–20,000 years and have adapted themselves as specialists (e.g., *Ascaris lumbricoides*) to the "late runner" on earth, i.e., the *Homo sapiens*. Parasites occur in many phyla of the Zoological Kingdom. Thus they are found among the unicellular organisms (Protozoa), among animals with a reduced number of cells (Myxosporida), among the various groups of worms (Plathyhelminthes, Nemathelminthes, Acanthocephala, Pentastomida, leeches), and among the billion-headed crowds of ticks, mites, and insects. The members of the latter three groups may also act as *vectors* of different pathogens (prions, viruses, bacteria, fungi, and parasites; see Tables 10.2, 10.3, 10.4, 10.5, 10.6). This transmission might be done just mechanically by means of contaminated mouthparts, by body contacts, and/or by placing pathogen-contaminated feces onto food, skin, or wounds. However, it may also occur as result of a defined developmental cycle with a reproduction of the pathogen inside the vector. Very often such reproduced pathogens even enter the eggs inside the female ovary or uterus so that the next vector generation is already carrier of the once ingested pathogen (i.e., this occurs inside ticks when the viruses of the spring-summer-meningoencephalitis or the kinetes (=motile zygotes) of the protozoan *Babesia* species enter the not yet fertilized eggs. Having entered the egg the parasites are firmly included therein and the thick eggshell, which is formed after fertilization and thickened during the laying process by adding a layer of wax, protects the growing tick larva and the therein included pathogens from desiccation. Examples of important zoonotic life cycles are compiled in Table 10.6) (Aspöck 2010; Darai et al. 2009; Getachew et al. 2007; Greenberg 1973; Kollaritsch and Paulke-Korinek 2010; Mehlhorn 2008; Mehlhorn et al. 2011; Mehlhorn 2012a, b, c; Oyerinde 1976; Szostakowska et al. 2004).

10.4 Food-Borne Zoonosis

When taking into consideration a very wide definition of the term *food-borne zoonosis* any pathogen that occurs inside or on meat of animals (inclusive fish) respectively is found within milk would fit into this group. However, in this case the differentiation from *contamination-borne zoonosis* would be rather difficult. Therefore it seems more reasonable to draw the limitations much stronger. This, however, leads to the fact that only a few groups of pathogens could be placed into this category. The prions—especially those of the bovine spongious encephalopathy (BSE)-group—would surely find their place here, since they introduce disease as well in animals as in humans after eating meat (even cooked one) of infected animals (Table 10.2). The viral pathogens with a zoonotic potential are mainly transmitted by bites of arthropods (mites, ticks, insects) or by fecal contamination, and only rarely—if at all—by infected food of humans and/or animals (Table 10.3). The explanation is that viruses need for their reproduction the very well-adapted DNA/RNA machinery of a living cell, which has problems to survive the passage of the stomach and intestinal system of humans (Altekrues et al. 1997; Darai et al. 2009; Neumeister et al. 2009; Steinmueller et al. 2006; WHO 2002a, b).

Likewise most bacteria prefer pathways of transmission other than the direct inclusion inside organs of warm- or cold-blooded animals respectively, fish. Thus there are not many bacteria like those of the *Brucella*-, *Listeria*- or *Salmonella*-groups that are frequently found not only on contaminated meat but also in milk (see Table 10.4).

Fungi with a zoonotic potential (Table 10.5) have no direct food-borne transmission pathways apart from fecal contaminations, that are common pathways in the species of the *Blastocystis*-group (if these species are really accepted as a fungus—see the other chapters).

On the other hand food-borne transmissions are very common in parasites (Table 10.6). There are protozoans such as several *Sarcocystis* species (Fig. 10.6) and *Toxoplasma gondii* which develop intracellular stages that later are transmitted, if this infected tissue is swallowed by a predator (inclusive humans). Then it starts a further developmental phase there. Man might act as *final host* (where the sexual process is running in his intestine, e.g., in *Sarcocystis* species) or as *intermediate host* with an exclusively asexual reproduction of the parasite in his tissues (e. g. in *Toxoplasma gondii*).

Also among the trematodes (phylum Plathyhelminthes) there are many examples of food-borne parasitosis in humans. In general humans are final hosts, since the adult worms live in their intestine (e.g., *Fasciolopsis buski*), in their lung (*Paragonimus* species), or in their liver (*Clonorchis sinensis*, *Opisthorchis* species, *Fasciola hepatica*, etc.; Table 10.6). The infection of humans with stages (metacercariae) of trematodes occurs by eating infected raw or undercooked intermediate hosts (e.g., fish, crustaceans) or plants with metacercariae that had been attached there by intermediate hosts (e.g., snails; Mehlhorn 2008).

For tapeworms (Cestodes; phylum Plathyhelminthes) humans might be final or intermediate hosts. In the case of the large human tapeworms *Taenia saginata*, *T. solium*, *T. asiaticum* and *Diphyllobothrium latum*, humans act as final hosts and shelter in their intestine the several meter long, hermaphroditic adult worms after being infected by ingestion of larval stages (cysticercus in the case of *Taenia* species respectively plerocercoids of *D. latum*) inside undercooked meat or fish. In the case of *T. solium* humans can be both final and intermediate hosts, since the cysticercus larvae may also develop inside their muscles and/or brain after they had orally ingested worm eggs from human feces (Mehlhorn 2008, 2012c).

On the other hand, however, in the cases of the tiny dog, respectively, fox tapeworms *Echinococcus granulosus* and *E. multilocularis* humans remain exclusively intermediate hosts, within which large cysts develop. Transmission of tapeworms of dog or fox to humans does only occur via the fecal contamination pathway when ingesting worm eggs from feces of dogs or foxes. However with respect to the general life cycle of these tapeworms humans are "dead ends," since in general they are not eaten by the final hosts (dogs, foxes).

Among the nematodes there are rather few examples for food-borne worm infections in humans with respect to the huge number of nematode species. However, one prominent example is given by the *Trichinella* species, of which up to now seven had been described. These worms live as adult, rather tiny worms (♀, ♂) of only 1–3 mm in length inside the intestine of predators (inclusive man) (Fig. 10.7). The females produce larvae; however, these stages do not leave this infected individual (i.e., acting here as final host), but they penetrate into the muscle cells, where they stay until another predator feeds such infected meat. Thus these predators are final and intermediate hosts at the same time and free stages do not occur outside a body (Mehlhorn 2008).

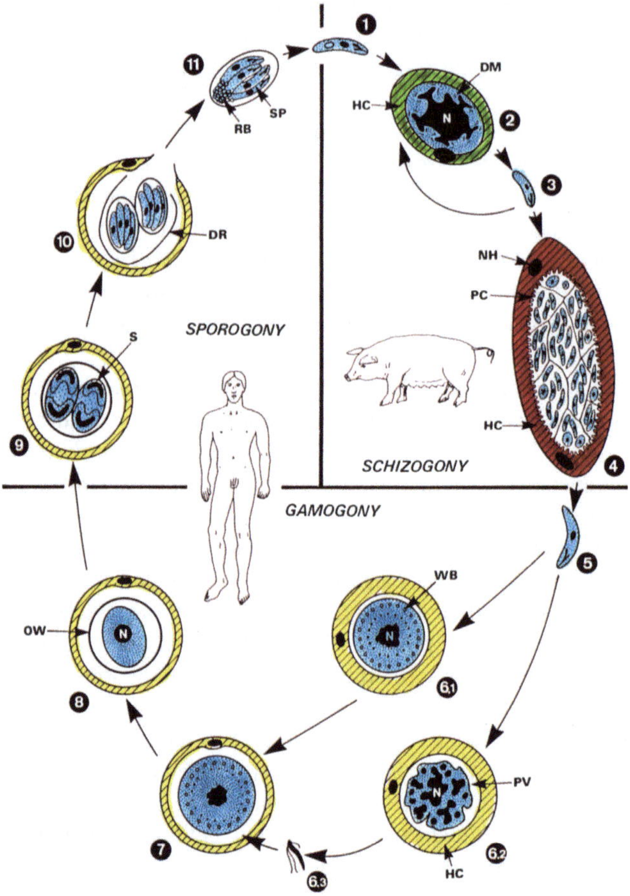

Fig. 10.6 Life cycle of *Sarcocystis suihominis* with two obligatory hosts. 1 Motile sporozoites hatch from the ingested sporocysts inside the intestine of the intermediate host, i.e., swine. 2 Two generations of schizonts are formed (5–6 and 12–17 days after infection) inside endothelial cells of blood vessels, giving rise to 50–100 merozoites by endodyogeny. 3 Free motile merozoites; first-generation merozoites enter other endothelial cells and form schizonts, whereas merozoites of the second generation induce formation of tissue cysts. 4 Cyst formation inside typical cells (muscle fibers, brain cells); within these cysts the parasites are reproduced by repeated endodyogeny leading to thousands of cyst merozoites which are situated inside chamber-like hollows. 5 When the final host man has eaten cyst-containing raw or insufficiently cooked meat, the cyst merozoites are set free and enter cells of the lamina propria. 6 Formation of female (macrogametes, 6.3) via gamonts (6.1, 6.2) within 14 h of infection. 7 Fusion of gametes. 8 Formation of the oocyst wall around the zygote. 9–11 Formation of two sporocysts (containing four sporozoites each) inside the host cell. The smooth oocyst wall often becomes disrupted. Thus, fully sporulated oocysts are found in the feces (11). *DM* developing merozoites; *DR* disrupted oocyst wall; *HC* host cell; *N* nucleus; *NH* nucleus of the host cell; *OW* oocyst wall; *PC* primary cyst wall; *PV* parasitophorous vacuole; *RB* residual body; *S* sporocyst; *SP* sporozoite; *WB* wall-forming bodies

Fig. 10.7 Light micrograph of a muscle fiber of a rat containing three larvae of *Trichinella spiralis*

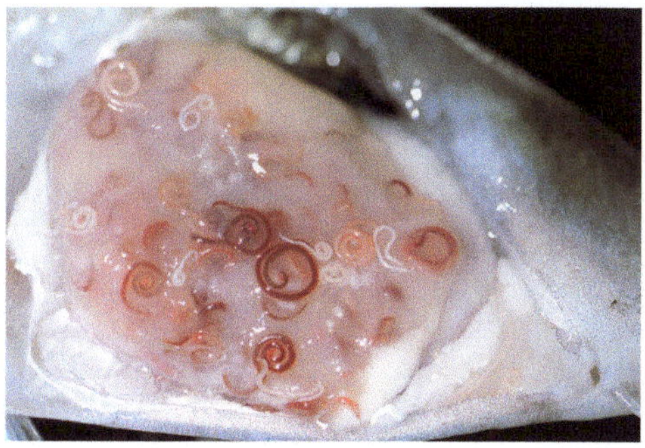

Fig. 10.8 Light micrograph of the muscles of a fish containing many *Anisakis* larvae

In the case of the species of the genera *Anisakis*, *Porrocaecum*, *Contracaecum* (Fig. 10.8), and related worms humans are intermediate hosts. Man, however, is not relevant for the transmission. He is a "dead end," since the adult worms live in the intestinal tractus of large marine mammals after they had ingested the larvae inside fish. Humans do the same: they eat raw fish containing these worm larvae, which live for a short time (1–3 weeks) in the human intestinal system before they die. Similar infections are also possible, when humans eat raw meat of reptiles; then they may ingest also larval pentastomids which finally may grow up to considerable size in humans (Mehlhorn 2008, 2012c).

10.5 Vector-Borne Zoonosis

Vectors in the sense of parasitological terms are animals that are able to transport pathogens from one host to another, while these vectors search their food on such hosts (Abbot 2002; Aspöck 2010; Hald et al. 2008; Holt et al. 2007; Löscher and Burchard 2010; Neumeister et al. 2009). The simplest way of such a transfer is the

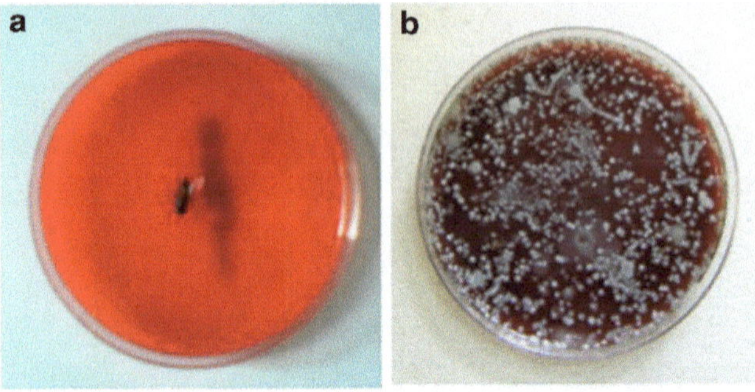

Fig. 10.9 Photograph of a petri dish filled with bacterial culture medium. Onto the surface of this petri dish a *Musca* fly was placed for 30 s (**a**). The photo at the *right* (**b**) shows the growth of bacteria 2 days after the exposition. This proves that the feet of this fly had been highly contaminated with bacteria

Fig. 10.10 Two adult *Ixodes* ticks laying eggs, which already may contain the viruses of the spring-summer meningoencephalitis

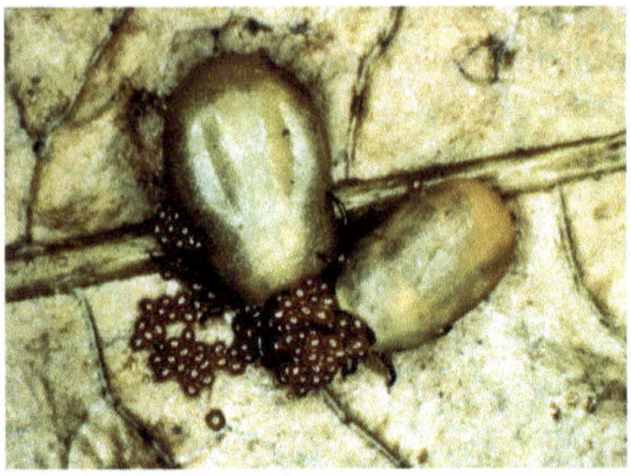

transport of pathogens that glue at mouthparts, at feet, or at the body surface of such vectors (Figs. 10.1, 10.2, 10.3, 10.9, 10.10, 10.11, 10.12), which are mostly mites, ticks, mosquitoes, flies, midges, simuliids, tabanids, leeches, etc. This type of transmission is described as mechanical, since the pathogens are not multiplicated on the vectors. For example, if 800 bacteria of a special species glue at the fine hair of a fly, the transmission maximum would be 800, but in general the finally transmitted amount is much less. Nevertheless this method might be still highly effective, if it is considered that many bacteria (such as e.g., toxic or hemorrhagic strains of *Escherichia coli* (EHEC, ETEC) need only an initial dose of 20–30 specimens to start severe infections. Recent studies examined various fly species of the genera *Musca*, *Calliphora*, *Sarcophaga*, *Lucilia*, etc. for their "normal load" of bacteria and parasites (protozoans and worm eggs). These flies had been caught close to stables of pigs, horses, cattle, rabbits, chickens, dog ponds respectively close to human recreation sites inside and outside of towns. Astonishing masses and varieties of pathogens were found on the examined flies. These studies of Förster et al. (2012) and Gestmann et al. (2012) confirmed earlier ones (Förster 2009; Förster et al. 2007, 2009) showing that more than 100 important bacteria species and more than 15 infectious stages of parasites might be transmitted via flies as vectors.

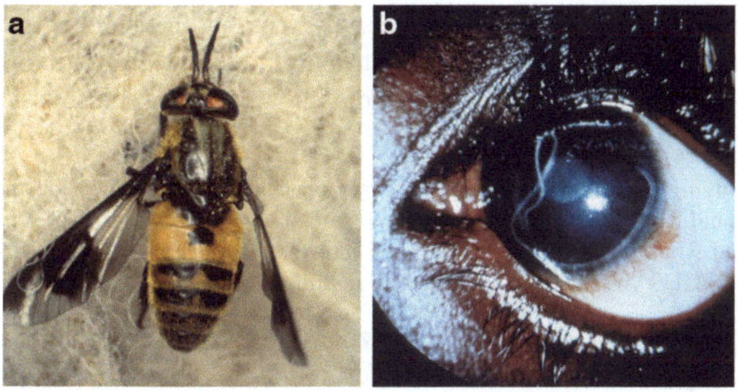

Fig. 10.11 (**a**, **b**) Photograph of a tabanid of the genus *Chrysops* and a wandering worm (*Loa loa*) in an eye (**b**, *right*)

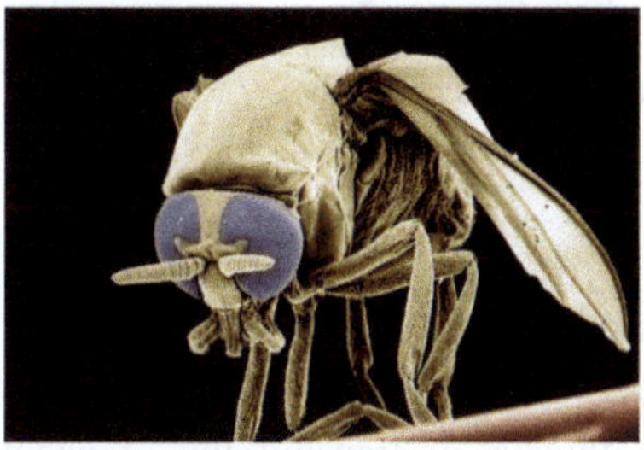

Fig. 10.12 Scanning electron micrograph of a black fly (*Simulium morsitans*), the vector of *Onchocerca volvulus* worms

Additional experimental transmission experiments showed that practically all types of bacteria were also transferred to bacterial culture plates, if the flies had contact for only 20–30 s to bacteria on a culture plate. In another series of experiments single flies were placed onto a fluid containing a definite number of *Toxocara canis* eggs. Then these flies were washed and squeezed (to get the intestinal fluid). When counting the obtained eggs from these flies the astonishing result was noted that merely two-thirds of the original number was found again.

These experiments clearly show that the importance of licking flies as mechanical vectors is apparently underestimated.

The same should be true for biting and/or blood sucking mites, ticks, mosquitoes, flies, fleas, lice, tabanids, etc., since most of them are able to switch quickly from one host to another, so that many tiny blood droplets remain liquid thus keeping the pathogens in an infectious status.

The situation becomes even worse in those cases, where a regular reproduction of the pathogens has been developed inside the vectors during evolution. Such reproductive processes occur regularly in vectors bearing special species of viruses, bacteria, and parasites (see Tables 10.3, 10.4, 10.6). Since these vectors even are able to transmit the pathogens into their eggs as e.g., the spring–summer–meningoencephalitis virus in ticks (Fig. 10.10), the distribution intensity is enormously enlarged not only by the flying activity of the vectors but also by the transmission into their progeny.

Therefore flying vectors that give pathogens a chance to reproduce inside their bodies represent a hidden risk, since they may help to spread severe epidemics. The consequences for mankind would have been terrific, if mosquitoes, tabanids, or ticks would be able to transmit the AIDS viruses as easy as they do it with other viruses.

Nevertheless the various vector-borne zoonosis demand even today a high death toll in many regions of the world. Considering the recent enormous increase of the Dengue-virus propagation by *Aedes* mosquitoes, the progress of the West-Nile-Virus into many countries as well as the still not yet solved malaria situation or filarial worm problems new and concentrated control measurements are highly needed. Especially in times of global warming and intense globalization epidemics may quickly become pandemics, which will not stop at the borders of countries, where the population believe even today to live on "safe grounds."

10.6 Cyclic Zoonosis

This term comprises pathways of transmission of agents of diseases that run repeatedly in a fixed direction and are based on at least two different hosts, where always defined developmental processes take place (Mehlhorn 2008). With respect to these conditions the transmission pathways of the malarial, babesial, and sarcosporidian parasites would fit in this category, because in these life cycles asexual processes like merogony (schizogony) start in a first host (human tissues) and are followed by gamogony and sporogony in a second host, which belongs—depending on the species—to the group of vectors (mosquitoes, ticks) or to mammalian species (as it is the case in the *Sarcocystis* species; Table 10.6, Fig. 10.6). Although the life cycle of the trematodes fits into this category, since, for example, the adult schistosomal worms live in the blood system of humans and other mammalians, while the asexual reproduction (via sporocysts and the finally infectious cercariae) takes places inside the tissues of water snails (Table 10.6). The life cycles of tapeworms belong to this group, too. Only if humans eat the *cysticercus-larva* with raw meat or the *plerocercoid-larva* in raw fish, the adult worms can develop from these ingested larval stages inside the human intestine (Table 10.6).

Among the nematodes examples of cyclic zoonosis are much more rare than in trematodes and cestodes (Aspöck 2010; Mehlhorn 2008). However, especially the filarial worms have developed very sophisticated developmental cycles, which need a very tight coordination in the different biotopes, if the transmission of the parasite from one host to the other should be successful. In the case of the filarial worms *Wuchereria bancrofti*, *Brugia malayi*, and *Loa loa* the adults live either in the lymph nodes or in the subcutaneous tissues (*Loa loa*), while their sheathed larvae

(=microfilariae enveloped by the egg shell) stay in the blood. However, since these three worm species use different vectors with different behaviors and activity periods, the larvae of these worms are concentrated at different day times in the peripheral blood vessels. *Wuchereria bancrofti* and *B. malayi* are transmitted by nightly active mosquitoes, thus their larvae are found in the peripheral blood mainly at 10 p.m. On the other hand *Loa loa* is transmitted by day time active tabanids (*Chrysops* species; Fig. 10.11). Thus their larvae are found in day-time from 1 p.m. to 3 p.m. in the peripheral blood vessels. The adult worms of *Onchocerca volvulus* (accumulations of their dead larvae in human eyes induce the so-called river blindness) stay in humans inside the subcutaneous tissues, where the females are often found in groups inside nodules. The infectious (unsheathed) larvae stay all day long in the lymph fluid of the skin system. There they are taken up by the simuliids (black flies; Fig. 10.12) which do not have as fine mouthparts as the mosquitoes, but scissor like ones, which lead to the formation of small "lacunes" of blood and lymph in the skin. Such an all-day-presence of the larvae in the skin makes sense for *Onchocerca* larvae, since the simuliids only suck blood during day time.

Besides all differences in their behavior inside humans the first larvae (=microfilariae) of all four species described above proceed a nearly identical development inside their vectors. Having reached the insect's gut the larva 1 enters the body cavity of the new host, grows up there (while molting twice), and the larva 3 enters finally the head of the vector. If this infected insect starts a new blood sucking action, the relatively thick larva 3 perforates the connecting membranes of the different mouthparts, creeps outside down to the biting site at the human skin and enters there the body of the new host.

Although this cyclic development seems very complicated, it is so perfect, that even today more than hundred fifty million humans in the tropics suffer from diseases due to these worms, since available means of chemotherapy are still rather poor.

10.7 Contamination-Borne Zoonosis

The diseases categorized inside this group are characterized by the fact that both humans and animals carry these pathogens, but their cross transmission occurs more or less occasional and the direction of the transmission is not fixed. This means that humans may get the pathogens from animals and that also the opposite way is possible (Tables 10.2, 10.3, 10.4, 10.5, 10.6). Examples for such contamination-borne zoonosis can be found in all groups of pathogens, which in addition also may use other pathways of transmission (WHO 2002a, b; Mehlhorn 2008).

Rabies viruses can be transmitted directly by saliva during bites of infected animals, by blood transfusion, or by contact to virus-contaminated saliva. Many bacteria change the host if there is contact either to human or animal feces or to contaminated food. The same is true for several species belonging to the group of fungi and thus also for the different *Blastocystis* species. Among the non cyclically transmitted parasites the species of the genera *Giardia* and *Cryptosporidium* as well

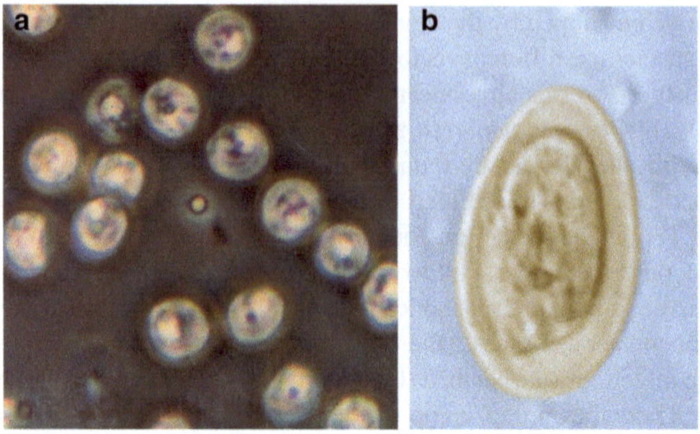

Fig. 10.13 Light micrographs of cysts of *Cryptosporidium* sp. oocysts (**a**) and a four-nucleated cyst of *Giardia duodenalis*. Such cysts had been excreted by infected animals in lakes in Milwaukee respectively Sydney. Humans became infected after drinking contaminated water and mass epidemics occurred

as the different microsporidian species (Table 10.6) belong to the group of contamination-borne zoonosis. In all these cases fecal stages (cysts, oocysts, respectively, spores) might contaminate food by direct contact with feces or by help of the transportation activity of licking flies (Table 10.6). But even in cyclically transmitted parasites such as *Trichinella spiralis*, *Plasmodium* species, *Trypanosoma* species, *Babesia* species, and *Leishmania* species mechanical transmission of the different pathogens may occur by contaminated mouthparts of large sized blood suckers (tabanids, leeches, blood licking bats) or even more frequently by transfusion of blood obtained from semi-immune donors with a low and thus only hardly detectible parasitemia.

Of course all these contamination-borne transmissions cannot become completely avoided—especially not in rural regions, however, strict hygienic conditions—as far as reasonable—will limit large outbreaks, as they had occurred in the USA or Australia when hundred thousands of humans had used drinking water that was contaminated with oocysts of *Cryptosporidium* species respectively with cysts of *G. lamblia* (*duodenalis*) (Fig. 10.13a, b).

Therefore preparation of safe drinking water will be worldwide one of the most prominent tasks of the future.

The outbreak of a kidney destroying disease due to hemorrhagic strains of *Escherichia coli* (EHEC, EPEC, etc.), which in 2011 took its start somewhere in Germany, spread due to travelers into many countries within weeks, and killed more than 300 persons, showed that the populations in crowded cities are highly vulnerable. Even today (i.e., 1½ years after the start of an intense search for the source of this epidemic), it is not completely finished and it remains unsolved, whether the finally detected contaminated soja sprouts had been the only source.

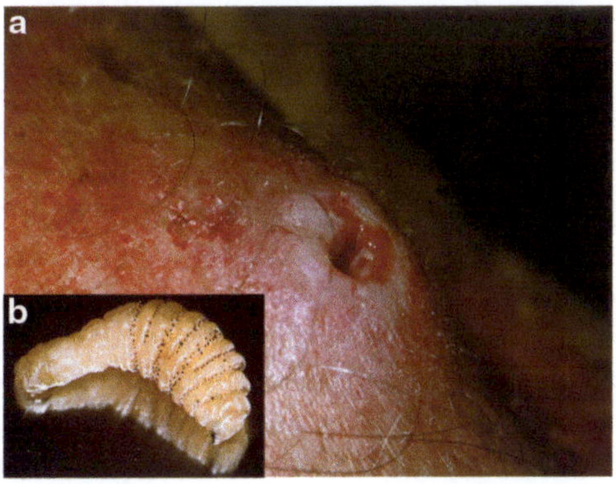

Fig. 10.14 Skin hollow (**a**) and the therefrom extracted second stage larva of *Dermatobia hominis* (**b**).

10.8 Myiasis

Myiasis, the name of which has its origin in the *Greek* term myia = fly, midge, is a very peculiar form of zoonosis (Mehlhorn 2008). Around the world there exists a group of flies, which do not proceed their larval development in debris of any kind or even not in dead bodies as it is done by many fly species, but which place their eggs or freshly hatched larvae onto the skin of animals and humans. These larvae bore—anterior end with the two mouth-hooks forward—into the skin. Exclusively the terminal end of the penetrated larva with its two terminal plates of the tracheal = oxygen uptaking system has contact to the surface. In order to avoid falling out unwillingly from their feeding hollow, these larva developed often very stiff circularly arranged hooks along their different segments. This peculiar appearance led to the description of these larvae as "screw worms." Some of these fly larvae are specialists and attack exclusively animals (e.g., *Haematobia bovis*). Others enter both humans and animals (e.g., *Oestrus* species), while again others have developed a peculiar preference for humans (*Dermatobia hominis*; Fig. 10.14).

Especially *D. hominis* has developed an unbelievable way of superinfection. The adult females catch while flying, a female mosquito of the blood sucking group and attach a bundle of own eggs on the body of this insect. As soon as this mosquito has a touch down on a host, the *Dermatobia* larvae hatch immediately during the sucking phase of their transporter mosquito and enter the skin—anterior end forward. After two molts the larva 3 finally leaves the inflamed hole in the skin, drops down to earth, and pupates there within a short time. Depending on the outside temperature the development of the adult stage inside the pupa takes only a few days, so that propagation can go on. In this case a broad spectrum of animals and the *Homo sapiens* are involved in a very effective, sophisticated manner. Thus these flies use mosquitoes and motile mammals (humans) to enlarge the spectrum of their biotopes.

10.9 Conclusions

The health of humans and their animals are threatened by a big bunch of pathogens, which have established during coevolution with their hosts a broad spectrum of pathways of transmissions. They all together belong to the group of zoonosis, where ping-pong infections occur between humans and animals. These infections might occur *directly* via contact to pathogen-containing feces, contaminated drinking water, or infected food. The *indirect* methods of transmission are based on the activity of so-called vectors, which transmit pathogens during blood sucking, while licking on skin and/or lips of humans or by contamination on human food.

With respect to these well-established pathways, the use and amelioration of measurements to avoid such transmissions is of high importance especially in increasing populations of animals and humans. Thus the following measurements have high priority in the near future:

- Preparation of safe, noncontaminated drinking water for humans and animals
- Quick clearance of human and animal feces
- To fight against possible vectors of pathogens in human houses and in stables of animals
- To use, wherever possible, vaccines and to develop new ones
- To establish a regular international control system to detect (at the very early moment) possible transmissions of pathogens and thus avoid outbreaks of epidemics and pandemics
- To keep masses of farmed animals in closed systems under most effective hygienic conditions
- To inform regularly farmers, leaders of institutions, veterinarians, physicians, and also the public on recent protection methods
- To intensify at the search for new compounds and new methods to control those pathogens, where treatment is not yet possible or where pathogens have developed resistances

References

Abbot SP (2002) Insects and other arthropods as agents of vector-dispersal in fungi. http://www.thermapure.com./pdf/AbbottInsectdispersal.pdf, 1–5 pp

Alam MJ, Zurek L (2004) Association of *Escherichia coli* O157:H17 with houseflies on a cattle farm. Appl Environ Microbiol 70:7578–7580

Altekrues SF, Cohen MT, Swerdlow DI (1997) Emerging foodborne diseases. Emerg Infect Dis 3:285–293

Aspöck H (ed) (2010) Sick through arthropods, vol 30, Denisia. Austrian Museal Publisher, Vienna

Banjo AD, Lawal OA, Adedujji OO (2005) Bacteria and fungi isolated from house fly (*Musca domestica* L.) larvae. Afr J Biotechnol 4:780–784

Bielaszewska M, Mellmann A, Zhand W, Köck R, Fruth A, Bauwens A, Peters G, Karch H (2011) Characterization of the *E. coli* strain associated with an outbreak of haemolytic uremic syndrome in Germany. The Lancet 11(9):671–676. doi:10.1016/S 1473-3099(11)70165-7

Darai G, Handermann M, Sonntag HG, Tidona CA, Zöller L (2009) Lexikon der Infektionskrankheiten des Menschen, 3rd edn. Springer, Heidelberg

De Jesus AJ, Olsen AR, Bryce JR, Whiting RC (2004) Quantitative contamination and transfer of *Escherichia coli* from food by houseflies, *Musca domestica* L. (Diptera: Muscidae). Int J Food Microbiol 93:259–262

Dobler G, Aspöck H (2008) Virus diseases. In: Mehlhorn H (ed) Encyclopedia of parasitology. Springer, New York

Dobler G, Aspöck H (2010) Krankheiten durch Viren. In: Aspöck H (ed) (2010) Sick through arthropods. Denisia 30:388 pp

Ekdahl K, Norman B, Andersson Y (2005) Could flies explain the elusive epidemiology of campylobacteriosis? BMC Infect Dis 5:1–11

Emerson PM, Lindsay SW, Alexander N, Bah M, Dibba SM, Faal HB, Lowe KO, McAdam KP, Ratcliffe AA, Walraven GE, Bailey RL (2004) Role of flies and provision of latrines in trachoma control: cluster-randomized controlled trial. Lancet 363:1093–1098

Fischer O, Mátlová L, Dvorská L, Svástová P, Bartl J, Melichárek I, Weston RT, Pavlik I (2001) Diptera as vectors of mycobacterial infections in cattle and pigs. Med Vet Entomol 15:208–211

Förster M (2009) Gefährliche Plagegeister—Rundschau Fleischyg. Lebensmittelüberwach 7:4–6

Förster M, Klimpel S, Sievert K (2009) The house fly (*Musca domestica*) as a potential vector of metazoan parasites caught in a pig-pen in Germany. Vet Parasitol 160:163–167

Förster M, Klimpel S, Mehlhorn H, Sievert K, Messler S, Pfeffer K (2007) Pilot study on synanthropic flies (e. g. *Musca*, *Sarcophaga*, *Calliphora*, *Fannia*, *Lucilia*, *Stomoxys*) as vectors of pathogenic microorganisms. Parasitol Res 101:243–246

Förster M, Gestmann F, Mehlhorn H, Sievert K, Messler S, Neuhausen N, Petersdorf S, Pfeffer K (2012) In: Mehlhorn H (ed) Flies as vectors of parasites potentially inducing severe diseases, vol 3, Parasitol research monographs. Springer, Heidelberg, New York

Fotedar R (2001) Vector potential of houseflies (*Musca domestica*) in the transmission of *Vibrio cholera* in India. Acta Trop 78:31–34

Fotedar R, Banerjee U, Samantray JC, Shirniwa S (1992) Vector potential of hospital houseflies with special reference to *Klebsiella* species. Epidemiol Infect 109:143–147

Gestmann F, Förster M, Mehlhorn H, Messler S, Neuhausen N, Pfeffer K (2012) Flies as vectors of parasites potentially inducing severe diseases in humans and animals. In: Mehlhorn H (ed) Arthropod as vectors of emerging diseases, vol 3, Parasitology research monographs. Springer, Heidelberg, New York

Getachew S, Gebre-Michael T, Erko B, Balkew M, Medhin G (2007) Non-biting cyclorrhaphan flies (Diptera) as carriers of intestinal human parasites in slum areas of Addis Ababa, Ethiopia. Acta Trop 103:186–194

Greenberg B (1973) Flies and disease, vol 2. Princeton University Press, Princeton, NJ, 447 pp

Grüntzig I, Mehlhorn H (2010) Expeditions into the empire of plaques. Düsseldorf University Press, Düsseldorf

Grüntzig J, Mehlhorn H (2010b) Robert Koch: Seuchenjäger und Nobelpreisträger. Spektrum Akadem Verlag, Heidelberg

Grübel P, Hoffmann JS, Chong FK, Burstein NA, Mepani C, Cave DR (1999) Vector potential of houseflies (*Musca domestica*) for *Helicobacter pylori*. J Clin Microbiol 35:1300–1303

Hald B, Skovgard H, Pedersen K, Bunkenborg H (2008) Influx insects as vectors for *Campylobacter jejuni* and *Campylobacter coli* in Danish broiler houses. Poult Sci 87:1428–1434

Holt PS, Geden CJ, Moore RW, Gast RK (2007) Isolation of *Salmonella enteritica* serovar enteritis from houseflies (*Musca domestica*) found in rooms containing *Salmonella* serovar enteritis-challenged hens. Appl Environ Microbiol 73:6030–6035

Karch H (2005) Enterohaemorrhagic *Escherichia coli* in human medicine. Int J Med Microbiol 295:405–418

Kobayashi M, Sasaki T, Saito N, Tamura K, Suzuki K, Watanabe H, Agui N (1999) House flies: not simple mechanical vectors of enterohaemorrhagic *Escherichia coli* O157:H7. Am J Trop Med Hyg 61:625–629

Kollaritsch H, Paulke-Korinek M (2010) Durchfallerkrankungen. In: Löscher T, Burchard KT (eds) Reisekrankheiten. Thieme, Stuttgart

Löscher T, Burchard GD (eds) (2010) Tropenmedizin in Klinik und Praxis. Thieme, Stuttgart

Mehlhorn B, Mehlhorn H, Walldorf V (2011) Schach den Blutsaugern und Schädlingen. Düsseldorf University Press, Düsseldorf

Mehlhorn H (ed) (2008) Encyclopaedia of parasitology, 3rd edn. Springer, Heidelberg, New York

Mehlhorn H (ed) (2012a) Arthropods as vectors of emerging diseases, 3rd edn, Parasitology research monographs. Springer, Heidelberg, New York

Mehlhorn H (2012b) Parasites of animals, 7th edn. Springer, Spektrum Akadem Verlag, Heidelberg

Mehlhorn H (2012c) Parasites of humans, 7th edn. Springer, Spektrum Akadem Verlag, Heidelberg

Mian LSH, Jacal JV (2002) Isolation of salmonellas from muscoid flies at commercial animal establishments in a Bernardino country, California. J Vector Ecol 27:82–85

Neumeister B, Geiss HK, Braun RW, Kimmig P (eds) (2009) Mikrobielle Diagnostik, 2nd edn. Thieme, Stuttgart

Nichols GL (2005) Fly transmission of *Campylobacter*. Emerg Infect Dis 11:361–364

Olsen AR, Hammack TS (2000) Isolation of *Salmonella* spp. from the housefly *Musca domestica* L. and the dump fly *Hydrotaea aenescens* (Wiedemann) (Diptera: Muscidae), at cage-layer houses. J Food Prot 63:958–960

Oyerinde JPO (1976) The role of the housefly (*Musca domestica*) in the dissemination of hookworm. Ann Trop Med Parasitol 70:455–462

Post K, Riesner D, Walldorf V, Mehlhorn H (1999) Fly larvae and pupae can serve as vectors for scrapie transmission. Lancet 84:1969–1970

Prusiner SB (1998) Prions. Proc Natl Acad Sci USA 95:13363–13383

Rabenau HF (2009) Diagnostik prionenbedingter Erkrankungen. In: Neumeister B et al (eds) Mikrobiologische Diagnostik. Thieme, Stuttgart

Rasko DA, Webster DR, Sahl IW et al (2011) Origins of the *E. coli* strain causing an outbreak of haemolytic uremic syndrome in Germany. New Engl J Med 365:709–717. doi:10.105/NEJMoa11069LO

Sasaki T, Kobayashi M, Agui N (2000) Epidemiological potential of excretion and regurgitation by *Musca domestica* (Diptera: Muscidae) in the dissemination of *Escherichia coli* O157:H/to food. J Med Entomol 37:945–949

Steinmueller N, Demma L, Bender JB, Eidson M, Angulo FJ (2006) Outbreaks of enteric disease associated with animal contact: not just a foodborne problem anymore. Emerg Infect Dis 43:1596–1602

Szalanski AL, Owens CB, McKay T, Steelman CD (2004) Detection of *Campylobacter* and *Escherichia coli* O157:H7 from filth flies by polymerase chain reaction. Med Vet Entomol 18:241–246

Szostakowska B, Kruminis-Lozowska W, Racewicz M, Knight R, Tamang L, Myjak P, Graczyk TK (2004) *Cryptosporidium parvum* and *Giardia lamblia* recovered from flies on a cattle farm and in a landfill. Appl Environ Microbiol 70:3742–3744

WHO (2002) Food safety and foodborne illness. World Health Organization (Fact sheet No. 237, Geneva)

WHO (2002) Foodborne diseases, emerging. World Health Organization, Geneva (Fact sheet No. 124)

Chapter 11
Conclusions

Heinz Mehlhorn, Minoru Yamada, Hisao Yoshikawa, Kevin S.W. Tan, Haris Mirza, and Kenneth Boorom

Abstract The most important aspects of the results obtained in the 101 years of research in *Blastocystis* are evaluated in this short chapter concluding the statement that this organism still remains an enigma, although all important methods had been used worldwide in this field.

Although huge progress has been achieved in the field of *Blastocystis* research since its discovery in the year 1911, this organism remained a sphinx with many controversial features. However, intense research using a variety of classic and most

H. Mehlhorn (✉)
Institute for Parasitology, Heinrich Heine University, Universitätsstr. 1, 40225 Düsseldorf, Germany
e-mail: mehlhorn@uni-duesseldorf.de

M. Yamada
Department of Infectious Diseases, Kyoto Prefectural University of Medicine, Kyoto, Japan

H. Yoshikawa
Department of Biological Sciences, Nara Women's University, Nara, Japan
e-mail: h.yoshikawa@cc.nara-wu.ac.jp

K.S.W. Tan
Laboratory of Molecular and Cellular Parasitology, Department of Microbiology, Yong Loo Lin School of Medicine, National University of Singapore, 5 Science Drive 2, Singapore 117597, Singapore
e-mail: mictank@nus.edu.sg

H. Mirza
Yale University School of Medicine, 333 Cedar Street, New Haven, CT 06510, USA
e-mail: Haris.mirza@yale.edu; Drmirza06@yahoo.com

K. Boorom
Blastocystis Research Foundation, 5060 SW Philomath Blvd, #202, Corvallis, OR 97333, USA
e-mail: kboorom@bhomcenter.org

H. Mehlhorn et al. (eds.), *Blastocystis: Pathogen or Passenger?*,
Parasitology Research Monographs 4,
DOI 10.1007/978-3-642-32738-4_11,

modern molecular biological methods has led to deeper insights during the last 20 years. Thus the following findings can be considered as widely confirmed.

1. *Blastocystis* stages are the most encountered microorganisms in the feces of a broad spectrum of animals and in humans.
2. Although there are species descriptions such as *Blastocystis hominis*, it is still doubtful whether there exist true species, since phylogenetic studies of the partial sequences of the SSU rRNA revealed at least nine subtypes. Thus *Blastocystis* belongs to the most genetically polymorphic unicellular organisms.
3. Traditional diagnostic methods when used commonly for inspections of probes of stool and/or animal feces showed that *Blastocystis* in general occurs as spherical globules with diameters varying between 5 and 50 μm.
4. When cultured in vitro also a size of up to 200 μm may be reached.
5. The stages obtained from fresh feces are characterized by a large central vacuole which pushes the cytoplasm with one to numerous nuclei, mitochondria-like organelles, Golgi apparatus at the peripheral rim. The central vacuole contains a broad spectrum of substances such as carbohydrates, lipids, and proteins being accumulated there by the activities of the Golgi apparatus and/or by crathrin-based endocytosis.
6. Due to their varying shape, the stages are described as vacuolar, multivacuolar, avacuolar, granular, amoeboid, or cystic. The vacuolar and granular forms are mainly seen in fresh feces or in in vitro cultures, while the mostly tiny amoeba-like stages are seen practically exclusively in cultures.
7. Locomotion or penetration of the intestinal wall has never been observed so that *Blastocystis* stages are typically inhabitants of the intestinal fluid.
8. The small *Blastocystis* stages in general contain one or two nuclei. If two occur, they were usually located at opposite poles of the cell. Larger stages contain several to many nuclei inside the small strand of peripheral cytoplasm.
9. Binary fission is apparently the most frequent reproduction seen as well in fresh stool samples as in in vitro cultures. However, also processes described as budding or plasmotomy were proven.
10. The actual mode of transmission of *Blastocystis* among the various hosts and/or the transmission from animals to humans and back is not yet conclusively demonstrated, although the occurrence of transmissions using apparently the fecal–oral route is very probable when looking at outbreaks among persons under tight living conditions or among animals in the same cage.
11. Experimental transmissions of human *Blastocystis* isolates to rats and the finding of human isolates in the intestine of chickens support the idea that there is a considerable zoonotic potential of *Blastocystis* stages.
12. The proposed oral–fecal route of transmission of *Blastocystis* is also supported by the fact that *Blastocystis* infections of humans are much more common in developing countries than in developed ones with higher sanitary standards. Especially contaminated drinking water seems of high importance in the transmission of *Blastocystis* stages.

13. The symptoms of disease, which are attributed to the occurrence of *Blastocystis* stages in the intestines of humans and animals, range between acute watery diarrhea, mild chronic abdominal discomfort until asymptomatic performances (being most common). However, there are also descriptions of urticarial and irritable bowel syndromes in humans, where *Blastocystis* stages are found in intestinal fluids.
14. *Blastocystis* has apparently a high opportunistic potential, since there are higher frequencies of *Blastocystis* in immunocompromised hosts including sick animals and pediatric and cancer patients as well as HIV-infected humans. There are also indications that the occurrence of *Blastocystis* stages in the intestine may lead to higher pathologic effects in parallel infections with invasive parasites.
15. It is suggested that the different clinical symptoms might be based on the occurrence of different subtypes, pheno- and genotypes of the parasitic *Blastocystis* stages observed.
16. Treatment of *Blastocystis* infection is poor—even today after a broad spectrum of reports using varying medicaments. Some groups recommend metronidazole—the medicament of choice against intestinal amoebae—as helpful, however, there are also a large number of reports of treatment failure when using this compound.
17. The systematic position of *Blastocystis* was discussed controversely for long, so that this peculiar organism wandered from the groups of flagellates to sporozoans, from there to amoebas (Sarcodina), to yeasts (*Saccharomyces* group), or fungi (*Neurospora* relatives). However, the recent investigations of the entire sequence of the SSU rRNA gene of human and guinea pig isolates of *Blastocystis* revealed clear hints that this organism apparently belongs to the group of stramenopiles. This group, which is also called Chromista or Heterokonta, is characterized by a unicellular body often with two differently sized flagella and contains species with photosynthesis like Xantophyceae (yellow-green algae) or colorless, flagella less groups such as Oomycetes, Opalinea, etc.

However, all findings indicate that the members of the genus *Blastocystis* belong to the most genetically polymorphic organisms among the wide spectrum of human and animal parasites. This may explain the numerous failures in treatment and/or the large variety of described subtypes. However, the last words are not yet spoken.

Index

H. Mehlhorn et al. (eds.), *Blastocystis: Pathogen or Passenger?*,
Parasitology Research Monographs 4,
DOI 10.1007/978-3-642-32738-4,

MIX
Papier aus verantwortungsvollen Quellen
Paper from responsible sources
FSC® C105338

FSC
www.fsc.org

If you have any concerns about our products,
you can contact us on
ProductSafety@springernature.com

In case Publisher is established outside the EU,
the EU authorized representative is:
Springer Nature Customer Service Center GmbH
Europaplatz 3, 69115 Heidelberg, Germany

Printed by Libri Plureos GmbH
in Hamburg, Germany